The Infamous Boundary

David Wick

The Infamous Boundary

Seven Decades of Heresy in Quantum Physics

with a mathematical appendix
by William Faris

COPERNICUS
AN IMPRINT OF SPRINGER-VERLAG

Paperbound edition published in 1996 by Copernicus, an imprint of Springer-Verlag New York, Inc., by arrangement with Birkhäuser Boston.

Published in the United States by Copernicus, an imprint of Springer-Verlag New York, Inc.

Copernicus
Springer-Verlag New York, Inc.
175 Fifth Avenue
New York, NY 10010
USA

Library of Congress Cataloging-in-Publication Data
Wick, David, 1950–
The infamous boundary : seven decades of heresy in quantum physics / David Wick ; with a mathematical appendix by William Faris.
p. cm.
Includes bibliographical references and index.
ISBN 0-387-94726-4 (softcover : alk. paper)
1. Quantum theory—History. I. Title.
QC173.98.W532 1996
530.1′2′09—dc20 96-18246

Manufactured in the United States of America.
Printed on acid-free paper.

9 8 7 6 5 4 3 2 1

ISBN 0-387-94726-4 SPIN 10533021

To my parents...

and to skeptics, heretics, and naïve realists everywhere.
Keep doubting; let others keep the faith.

Acknowledgements

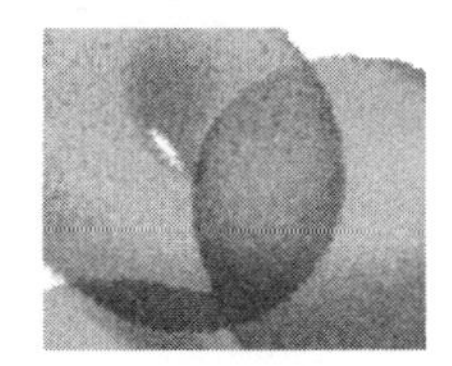

A great number of people assisted the author during the years he actively wrote the text; many more, too many to cite, taught him quantum mechanics and philosophy of science in previous decades. Apologies are in order to those whose ideas were altered, abused, or simply filched; after all, that is a writer's basic strategy.

Henry Atha, Lee Anne Bowie, Stephan Faris, Chris Kirschner, Joe Mabel, Tim Wallstrom, John Westwater, and Matthew Howard each read a part or the whole of a draft and made useful comments. Arthur Wightman and Abner Shimony provided lengthy critiques of a late draft, enabling the author to avoid the worst gaffs; any remaining absurdities are not their fault. (Special thanks to Arthur for agreeing to review a "heretical" book when he had more productive ways to use his time.) I thank John Clauser, Hans Dehmelt, Stuart Freedman, Richard Holt, Michael Horne, Werner Neuhauser, Abner Shimony, and their collaborators for interviews, and for providing unpublished or original materials. These scientists were generous with their time to an unknown author. I thank Matthew Howard and Blaine Walgren for pointing out obscure references, and anonymous reviewers for fixing the odd factual mistake.

I thank my thesis advisor, John Westwater, and the Department of Mathematics of the University of Washington for their support (including, but not limited to, providing that all-important Internet access) during this project.

I thank Lee Anne Bowie, Eddie Lock, Jordan Lock, and Elizabeth Lock for enduring the author's presence as he wrote the first draft of this book in their dining room. (To Jordan and Elizabeth: sorry, but the parable of the two children and the candy store did not make it in.)

I thank Charles Newman for reading a draft and helping arrange publication.

Finally, special thanks to my friend and colleague William Faris for helping to define the audience; reading and critiquing multiple drafts; suggesting a title; and encouraging, arguing with, and cajoling the author into finishing the project, especially when he had given it up as beyond his abilities. I thank Bill also for writing the Appendix contributing some genuine mathematics; whatever can be said of the first part of this book, the last can be recommended unreservedly. I thank the University of Arizona, Department of Mathematics, Bill Faris and Madeleine Lapointe for their hospitality on my visit to Tucson, and Madeleine for a wonderful anecdote.

Contents

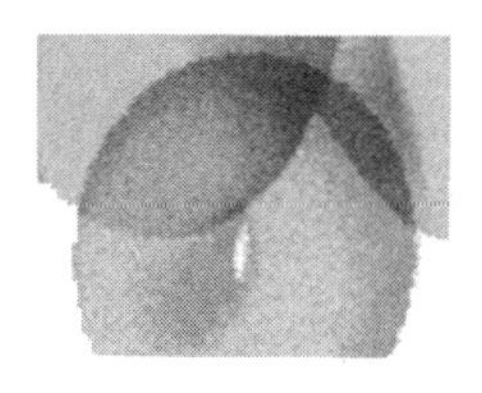

Introduction

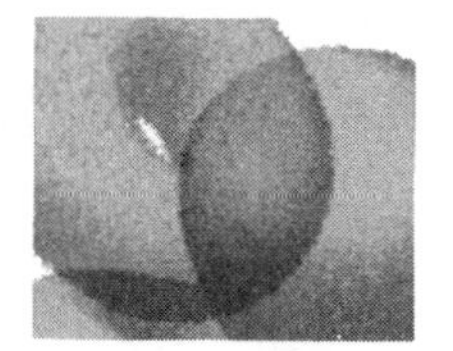

On an October morning in 1927, most of the world's atomic experts had gathered at the Hôtel Metropole in Brussels for a conference. Among the senior physicists present were Niels Bohr of Denmark, Max Born and Albert Einstein of Germany, and Erwin Schrödinger of Austria; the younger scientists included Louis de Broglie of France and Werner Heisenberg of Germany. The conference topic was an atomic theory that had been invented independently, and in very different guises, by Heisenberg and Schrödinger a little over a year before. Many conference participants hoped it would resolve the "wave-particle paradox" that had haunted physics since the turn of the century—but the proper interpretation of the new theory was far from obvious.

De Broglie spoke first, outlining a simple, realistic scheme based on a wave picture of Schrödinger's, meant to banish the paradox. The younger theorists derided this proposal, although Einstein said he thought de Broglie was on the right track. Then Heisenberg and Born (Heisenberg's colleague and mentor in Göttingen) gave a joint talk in which they advocated a radically different interpretation and at one point declared: "... quantum mechanics is a complete theory; its basic physical and mathematical hypotheses are not further susceptible to modification."

In the terminology of the Göttingen physicists, a "quantum mechanics" meant any new theory of particles (electrons and protons) that would explain the peculiarly discrete or "quantal" properties observed on the atomic scale. Thus everyone understood Born and Heisenberg's statement to be, besides a repudiation of de Broglie's proposal, a declaration of victory over the unruly atom—a triumphant cry after 20 years of mostly disappointing efforts by such preeminent theorists as Bohr and Einstein.

It was the climactic moment of a scientific revolution—yet the paradox remained. In the afternoon Niels Bohr advanced a new scientific philosophy embracing contradictory pictures of a single phenomenon as a virtue. Einstein objected, and a debate ensued. It continued the next day at breakfast and lasted into the evening. At subsequent meetings it broke out again and then led to clashes in the scientific journals. The mealtime dialogue had developed into a grand intellectual contest, which lasted nearly three decades. Although the establishment awarded the decision to Bohr (and Einstein's followers became increasingly isolated and marginalized), Einstein insisted until his death in 1955 that quantum mechanics gave a useful but *fundamentally incomplete* account of the physical world.

Today most physicists accept quantum mechanics as the basis of their discipline. And as physics lays claim to the foundation stones, it may be the ultimate basis of all the sciences. Yet, strangely, the controversy that started in a hotel lobby six decades ago is no mere historical footnote.

Consider the views of the late Irish physicist John Bell. In a speech before a professional audience in 1989, Bell charged that physicists had divided the world into two realms—a "classical" one and a "quantum" one—with no intention of explaining what happens at the boundary between them. (Bell gave his talk the provocative title "Against 'measurement'," the meaning of which will become clear in Chapter 18, "Principles.") Bell's opinion carried special weight, since in 1964 he proved a theorem about quantum mechanics that has been called the most surprising discovery made by a physicist in this century. (See Chapter 11, "Bell's Theorem.") But Bell (who died prematurely in 1990) was not alone in his critique. Many others in the quantum-orthodoxy-doubting subculture, from the founders Einstein and Schrödinger to the American theorist David Bohm (a contemporary and rival of Richard Feynman, see Chapter 10, "The Post-War Heresies") and the American experimentalist John Clauser (Chapter 13, "Testing Bell"), spoke out about the ruling paradigm.

These physicists did not claim that the emperor has no clothes. Indeed, they acknowledged that in its physical predictions quantum theory has been a great success. Quantum mechanics correctly described the color of gases, the heat capacity of solids, and the nature of the chemical bond; it explained the periods in the chemists' Periodic Table of the Elements; it calculated the electrical properties of conductors, semiconductors, and insulators; it provided a theoretical foundation for lasers and masers, superfluids and superconductors; it presaged electron imaging and neutron diffraction, by which the structure of a host of materials was discovered; it established many properties of radioactivity and of the elementary particles; and there were other successes. Nevertheless, argued these physicists, certain vital areas of the royal anatomy were covered, not by a stout weave of convincing computations, but by a transparent tissue of ideology.

The monarch's courtiers, of course, have not been silent; their voices are frequently heard decrying these heretical views. The ongoing debate is rancorous and voluminous. Hundreds of journal articles, many texts, monographs, conference proceedings, and even popular books are devoted to the controversy every year. New solutions to the quantum paradoxes, some barely this side of lunacy, appear with regularity. Apologies for Bohr's dualistic philosophy can be read side by side with encomiums on Einstein's realism. When Bell's remarks were reprinted in the British trade journal *Physics World* in 1990, three separate and lengthy replies from establishment physicists were printed in subsequent issues. (See Chapter 18, "Principles." Similar outbreaks occur regularly on this side of the Atlantic, in *Physics Today* and other journals; see Chapters 14, "Loopholes," and 20, "Speculations.")

For outsiders, especially scientists who rely on physicist's theories in their own fields, this situation is disquieting. Moreover, many recall their introduction to quantum mechanics as a startling, if not shocking, experience.

A molecular biologist related how he had started in theoretical physics but, after hearing the ideology of quantum mechanics, marched straight to the Registrar's office and switched fields. A colleague recalled how her undergraduate chemistry professor religiously entertained queries from the class—until one day he began with the words: "No questions will be permitted on today's lecture." The topic, of course, was quantum mechanics. My father, an organic chemist at a Midwestern university, also had to give that dreaded annual lecture. Around age 16, I picked up a little book he used to prepare and was perplexed by the author's tone, which seemed apologetic to the point of pleading. It was my first brush with the quantum theory.

Eventually, I went to graduate school in physics (but switched quickly to mathematics, albeit to a branch called "mathematical physics"; the interesting ideological distinction is drawn in Chapter 1, "Prologue I: Atoms"). By then I had acquired a historical bent, which developed out of an episode in my freshman year in college. To relieve the tedium of the introductory physics course, I set out to understand Einstein's theory of relativity (the so-called Special Theory of 1905, not the later and more difficult General Theory of 1915). This went badly at first. The textbooks all began with some seemingly preposterous proposition such as the universe having a speed limit, or a light wave's apparent velocity not changing if the observer goes into motion. Had Einstein never passed a boat's wake while sailing on that pond in Princeton? I felt like Alice advised by the White Queen to practice her credulity: "Why, sometimes I've believed as many as six impossible things before breakfast." Like Alice, I couldn't manage it.

Prozac being unavailable at the time, I was saved from depression by a little book I found at a Hyde Park bookstore containing translations of the original articles by Einstein, Minkowski, and others. Here were the arguments that had convinced physicists to abandon the world view of the great Isaac Newton for that of a patent clerk. Over Christmas break I read Einstein's first paper, and light instantly banished the fog of incomprehension. (See Chapter 18, "Principles," for Einstein's simple axioms, which had become garbled in the minds of textbook writers.)

Four years later and bewildered again, this time by quantum theory, I seized upon a translation of an early monograph by one of the theory's founders, hoping for a similar epiphany. But, to my amazement, modern textbooks had reproduced the original arguments. As I manipulated the infinite matrices and imaginary quantities of the quantum theory, I could be heard muttering: "Someday I am going to understand this crazy business."

I retreated to a related field (statistical mechanics, in which probabilistic and mechanical arguments are combined to describe the various states of matter; see Chapter 1, "Prologue I: Atoms") but spent many hours (usually after midnight) absorbing the quantum mechanics literature. At Princeton and elsewhere I sought out heretics as well as establishment figures, and I asked their views. Many opinions that I heard in those years—especially from orthodox physicists—were strange indeed. Twenty years passed. Some paradoxes were resolved (while others increased in importance), and experimentalists attained remarkable powers over elementary particles. (Among other accomplishments, a scientist became the first in human history to glimpse a solitary atom with the naked eye; see Chapter 15, "The Impossible Observed.") But the counterrevolution did not come. Declining to wait for that event, I decided to commit what I had learned to paper (or, more accurately these days, to disk).

In this book I have tried to give almost equal time to orthodox physicists and to heretics—to the believers and to the skeptics. Thus it is primarily the history of an intellectual struggle. Because a comprehensive treatment would have run to many volumes, I was forced to be ruthless in selecting topics. References to more detailed histories or biographies, where available, are found in the Notes and in the Bibliography. The book is arranged chronologically, beginning with the battle over the reality of atoms that preoccupied physicists at the end of the last century.

I address this book to scientists—at least those who can tolerate the lack of technical jargon and equations—and to interested laypersons. For the most part, I stick to the science. I abhor the tendency, so evident recently, to relate every difficulty scientists are experiencing to Buddhism, Christianity, the occult, or other mystical traditions—as if modern science, after a mere four centuries, had attained its final limits and therefore any remaining mysteries must relate to Something Outside. (For example, a man recently packed a theater in my town by announcing a lecture on "Quantum Mechanics and God.") Einstein, Bohr, Heisenberg, Schrödinger, and Bell all regarded the dispute as primarily a *professional matter*, and I see it the same way. (Niels Bohr did think the quantum paradoxes reflected a more general limitation on scientific thought, but even he did not link them directly to mysticism.)

This said, I hasten to remark that scientists do have philosophies, and they can make a difference. Indeed, the quantum controversy is a contemporary expression of a philosophical dispute that occasionally has divided scientists into hostile camps. The currently accepted labels for the opposed positions are "realism" (often modified by "critical" or "naïve," depending on the writer's persuasion), describing Einstein's, Schrödinger's, and Bell's views; and "positivism," for Bohr's, Heisenberg's, and that of the majority of physicists today. Many people recognize Einstein's aphorism "God does not play dice" and even know that he aimed it at quantum theory. He did say that, but it was not his primary objection. What truly riled him was that physicists had adopted an antirealistic world view based on subjectivism and "complementary thought." In a more revealing moment, Einstein once turned to the physicist Abraham Pais during a walk in Princeton and asked, "Do you really think the moon is only there when you look at it?" Philosophy mattered to these scientists, and so it must concern us as well. (To satisfy the demands of openness and intellectual honesty, I admit here to being a naïve realist, as defined in Chapter 17, "Philosophies." I expound my own views in the final chapters.)

Since I am not a professional philosopher but a practicing scientist, I make no pretense of having discovered the proper context in which to discuss these philosophies. Those interested in locating, for example, Bohr's antirealism in a philosophical tradition going back to Hegel, Kant, or (perhaps) Plato will be disappointed. Such treatises can be found in the literature; a few are listed in the Bibliography.

Of the great scientific controversies of this century, the battle over quantum mechanics involved the widest span of scientific issues, as well as the subtlest arguments. I will try to smooth out the rough parts where I can, but, unavoidably, some topics will tax the reader's intuition, just as they did the author's (not to mention the thousands of investigators who preceded us). As to technical matters, equations other than innocent proportionalities will be eschewed in favor of prose summaries, except occasionally in the Notes. (I make one exception in Chapter 4, "Revolution, Part II: Schrödinger's Waves," for an equation whose bizarre typography, like the arrangement of the saints in medieval paintings, embodies part of the mystery.) However, Professor William Faris of the University of Arizona, an expert on the mathematics of quantum mechanics, has written a brief introduction to the theory which requires no matrix algebra or even calculus to understand. It is included as an appendix and will be referred to occasionally in the text; I recommend it highly to anyone who does not experience palpatations at the sight of a solitary Greek letter.

Ninety years ago, one young scientist's questioning of the foundations of his field lead to the formula $E = Mc^2$; the false dawn at Alamogordo and the nuclear power industry were eventual consequences. Could today's battle over principles in quantum physics have a similar impact? Most physicists regard quantum mechanics as the fundamental theory of matter and subscribe to the founders' view that it is a complete theory of physical reality. (How they are able to do this while ignoring gravity will be the subject of some comments in the last chapter, "Speculations.") But they may be wrong. Perhaps even as I write, some young scientist has found a radically new way to describe the microworld inhabited by elementary particles, a way that will extend easily and without paradox to the macroworld inhabited by human beings. Who can guess what might then become possible?

1.

Prologue I: Atoms

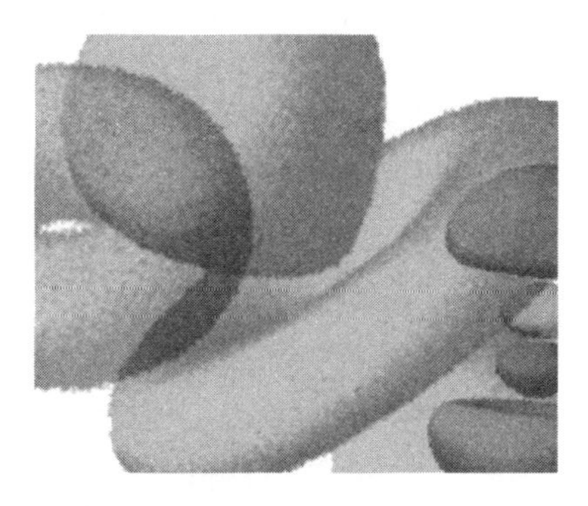

Now that a visitor to a physics laboratory can be shown a solitary barium ion glowing like a tiny star caught in a bottle, skepticism about the existence of atoms seems as quaint and remote from our time as belief in a flat Earth. Yet less than a century ago, serious thinkers still doubted the reality of atoms.

In 1895, Ernst Mach, physicist, psychologist, and philosopher, returned after a long and involuntary absence to his *alma mater*, the University of Vienna, to become *Ordinarius* Professor "for the history and theory of the inductive sciences." Mach—although previously little known in Austria—quickly became the darling of the cognoscenti in the Ring City. His views on scientific philosophy and psychology were widely discussed, and his Buddhistic utterance *Das Ich ist unrettbar*—"the I is unsavable"—even became a catchphrase in the city's coffeehouses.

Among physicists, however, Mach's return created a perplexing situation. A reminiscence of Ludwig Boltzmann, professor of physics in Vienna after 1894, illustrates the dilemma confronting the physics student in *fin de siècle* Vienna:

> I once [1897?] engaged in a lively debate on the value of atomic theories with a group of academicians, including Hofrat Professor Mach, right

> on the floor of the academy itself.... Suddenly Mach spoke out from the group and laconically said: "I don't believe that atoms exist."

Mach's remark derived its punch from the recent history of Viennese physics. Following Mach's departure three decades before, Vienna had become a world center of atomism. First Josef Stefan, who had defeated Mach in 1864 in a contest to become head of the Physical Institute, then Josef Loschmidt, and finally Boltzmann had championed the atomic theory in Vienna. Loschmidt in particular was the first to estimate both the number of molecules in a fixed volume of gas (roughly 10^{19} in a cubic centimeter at zero degrees Celsius and one atmosphere pressure) and the size of a molecule (about 10^{-8} cm). Loschmidt obtained his estimate from James Clerk Maxwell's "kinetic theory" of gases (1859), which attributed the pressure of a gas in a vessel to the continuous bombardment of the vessel's walls by the gas molecules, and the temperature to the energy of their motion.

The Italian chemist Avogadro had conjectured earlier that equal volumes of gases at the same temperature and pressure contained equal numbers of molecules, but he could not compute their numbers. The chemists, as everyone recalls from his or her school days, preceded the physicists in formulating the modern atomic theory. (By "modern" I mean as opposed to classical: "atom" was derived from the Greek *atomos*, meaning indivisible, and the honor for introducing atomism into philosophy traditionally goes to Democritus of Abdera, who lived around 430 B.C., although Hindu philosophers had similar ideas.) The French chemist Antoine Lavoisier's law of conservation of matter (1789) and his countryman Joseph Proust's law of fixed proportions in chemical reactions (1797) had made the atomic hypothesis almost unavoidable; in 1803, the English chemist John Dalton put forward the full conception: elements are formed from identical atoms with equal masses, and molecules are formed from unions of these atoms. (Dalton thought the integer ratios of atomic weights proved the theory, and he maintained his faith even after chemists made more careful measurements and discovered some ratios were not integral, e.g., chlorine/hydrogen = 35.5, which proves the value of stubbornness when confronted by unruly facts.)

After the premature death of Maxwell in 1879, Boltzmann became the leading proponent of the kinetic theory in Europe. Five years earlier, Boltzmann had penned an equation that described how collisions in a dilute gas drive the gas toward a state of equilibrium. It was the first mathematical proof of the second law of thermodynamics—the celebrated rule that entropy, a measure of randomness, in a closed system with a "sufficiently chaotic" initial state always increases. Boltzmann and the great American theoretical physicist Josiah Willard Gibbs, of New Haven, Connecticut, co-founded the field now called "statistical mechanics," in which dynamical laws are combined with statistical assumptions in an attempt to understand gases, liquids, and crystals. By the time Mach returned to Vienna, Boltzmann was widely hailed as Europe's greatest living theoretical physicist.

Mach attained fame among physicists both as a critical thinker (he would have cringed at the word "theoretician") and as an experimentalist. Most people have heard of "Mach numbers," which measure velocity as a multiple of the speed of sound; Mach was the first to study the shock wave produced by a bullet when it exceeds that velocity, and he made other equally meritorious contributions to experimental physics. One other scientific work of Mach's should be mentioned: the *Science of Mechanics* (1883), in which he made a devastating criticism of Newton's concept of absolute space. Thirty years later, Einstein made use of Mach's analysis in formulating his general theory of relativity. (Strangely, despite the "Machian" element in relativity theory, Mach opposed it.)

Boltzmann's surprise at Mach's laconic interjection was surely feigned, for he had been troubled from afar by Mach's philosophical views for more than a decade. Although Mach's scientific philosophy resembled the "positivism" of Auguste Comte (who also gave us "sociology"), the primary influence on Mach's philosophy was the science of thermodynamics. This discipline treated the relations among temperature, pressure, and other directly measurable quantities without invoking fictitious "atoms." Mach answered the challenge of epistemology—what we can know?—and ontology—what exists?—with a single word: *sensations*. Primary sensations alone are knowable and universal, Mach argued; all the rest is speculation. The goal of science is simply to relate the primary sense impressions (Mach called them "elements") to each other in the most economic way. By this Mach meant by the most economic mathematics: Mach admired mathematical physics but despised *theoretical* physics. (Nowadays physicists tend to have these sentiments reversed. One recalls Richard Feynman's remark that if all of mathematics disappeared, physics would be set back by exactly one week—to which the probabilist Mark Kac retorted: "Precisely the week in which God created the world.")

When people would bring up atoms, Mach would ask, "Have you seen one?" Followers expanded on his skepticism. Wilhelm Ostwald pushed a positivistic theory called "energeticism" until he lost a public encounter with Boltzmann. (In 1892, Ostwald had performed the amazing feat of writing a chemistry textbook that failed to mention atoms or molecules.) Philosophers, especially the Vienna Circle in the 1920s, absorbed Mach's phenomenalism and combined it with symbolic logic and other modern developments into the system called "logical positivism," which still has adherents today.

No single phrase adequately describes Boltzmann's scientific philosophy, but "conjectural realism" comes closest. ("Naïve realism" is sometimes employed, usually by Machists with a sneer. The terms "positivism" and "realism" will occur throughout this book; I define them more precisely in Chapter 17. Although we are stuck with these terms for historical reasons, they are peculiar choices: positivism is actually negativism about scientist's speculative creations, while realists are people willing to believe in things they cannot see!) Boltzmann emphasized the role of speculative hypotheses, arbitrary inventions of the human mind, in scientific theories. He thought science without hypotheses not even desirable—"the bolder one is in transcending experience, the greater the chance to make really surprising discoveries... the phenomenological account... has no reason to be so proud of sticking so closely to the facts!" But lacking direct proofs, ultimately Boltzmann could only imitate Galileo in his debate with Cardinal Bellarmine three centuries before. The Church insisted that the Earth was at rest and that the Copernican theory was a mere calculating tool, but Galileo, under duress, supposedly muttered,

"It moves!" What could Boltzmann further say in defense of atoms but "They move"?

Mach suffered a stroke in 1898, which paralyzed the right side of his body and forced his retirement in 1901. Nevertheless, he survived long enough into the 20th century to fight battles with Planck and Einstein over positivism and realism. When shown alpha rays scintillating on a screen in 1903, Mach said jokingly "Now I believe in atoms," but he remained stubborn until the end, writing a year before his death in 1916, "... in my old age, I can accept the theory of relativity just as little as I can accept the existence of atoms and other such dogma."

After Mach retired, Boltzmann was appointed to his chair in philosophy—a magnetic reversal for Viennese physics. Boltzmann might have considered it a personal triumph; indeed, among the literati of Vienna, Boltzmann soon enjoyed the popularity that Mach once had. But the positivists continued their assault, disparaging Boltzmann as "the last pillar" of atomism. In 1906, while on summer vacation in Italy, Boltzmann hanged himself. It was asserted that the attacks by the Machists contributed to Boltzmann's self-annihilation, but in his last months Boltzmann suffered from severe headaches, failing eyesight, and a "nervous condition" that might now be diagnosed as depression and treated with drugs. Thus the truth of the accusation can never be known.

One effect of the rejection of atoms by Mach's school was to compel the next generation of physicists to find definite proof of their existence. Albert Einstein, Max Planck, Marian Smoluchowski, and others took up the task, and by around 1910, a dozen independent determinations of Avogadro's (or Loschmidt's) number were known, all in agreement. The reality of atoms could no longer be denied, and adherence to Mach's philosophy among physical scientists declined accordingly.

This eclipse of positivism lasted a mere 15 years: Mach's philosophy re-emerged full-blooded in the first sentence of the first paper on quantum mechanics.

2.

Prologue II: Quanta

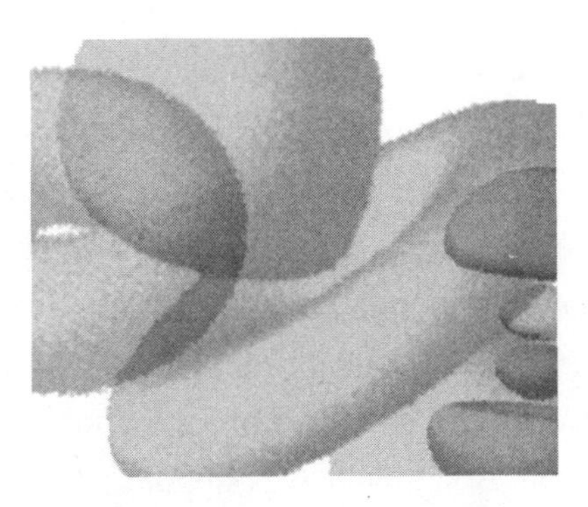

As the nineteenth century came to a close, physicists were in an upbeat mood. The discoveries of Roentgen, Becquerel, and the Curies—x-rays, penetrating radiation, and the transmutation of elements—had capped a triumphant century in the laboratory. On the theory side, Hamilton, Jacobi, and others had solved the problems of rigid bodies and constrained motions, completing Newton's program in mechanics. Faraday and Maxwell had unified electricity, magnetism, and optics at one stroke in their continuum theory of the "aether." True, these two great edifices were somewhat askew at the join, and the building labeled "atomic structure" remained an empty shell. But only a few physicists worried about that.

A senior scientist judged the result, remarking that since all the basic theories were known, the primary task of physicists in the future would be to make ever-more refined measurements of the fundamental constants. Yet a mere five years would pass before a clerk at a Swiss patent office mailed papers to a German journal suggesting that neither house of nineteenth-century theoretical physics rested on firm foundations. As for atomic structure, hopes for a quick solution would be dashed almost before physicists had completed their year-end salutations and gone back to the laboratories.

In 1899, the English physicist J. J. Thomson demonstrated that certain rays seen in electric discharges were actually streams of minute corpuscles, each carrying the identical amount of negative electric charge. A term proposed earlier by the Irish physicist George Stoney for a hypothetical discrete unit of charge, the "electron," was adopted for the new particle. (The obscurity of this scientific field at the time is perfectly captured by the toast once offered in Cambridge: "The electron—may it never be of any use to anyone!") Theorists immediately speculated that the atom was made up of electrons. This was only natural, but as atoms are normally found in an electrically neutral state, positive electricity had also to reside somewhere in the atom. Thomson favored a uniform background of positive charge in which the electrons swam like minnows in a pond; others proposed that another particle, not yet discovered, carried the positive charge. There was even speculation, anticipating Rutherford by a decade, that the atom was a miniature solar system, with a positively charged central "sun" around which the negatively charged "planets" (electrons) revolved.

These proposals had to surmount two obstacles to atomic modeling—one theoretical and one experimental. The theoretical difficulty concerned stability. Before treating any of the transformations to which atoms are prone, theorists had first to explain the relative permanence of atoms. (By way of analogy, a historian who sets out to explain why an empire of a thousand years finally collapsed should explain first why it survived for so long.) A theorem from electrostatics stated that no system of stationary charged particles could be stable. Therefore, the electrons had to move. Obviously they could not move in straight lines, since the atom would fly apart. So their trajectories must be curved, but now Maxwell's theory of the electromagnetic field caused headaches. A particle moving in a curved orbit is accelerated (according to Newton) and, if charged, must continuously radiate electromagnetic waves (according to Maxwell). But electrons in atoms did nothing of the sort: atoms were observed to persist in stable states, without radiating, for indefinite periods.

The experimental conundrum concerned the reverse problem: how atoms emitted or absorbed radiant energy such as light. By the middle of the century, physicists had abandoned Newton's corpuscular theory of light propagation for Young's and Fresnel's wave theory. Each color of light was thought to correspond to a definite frequency of vibration and to a corresponding wavelength. In 1853, the Swedish scientist Ångström observed emission "lines" from hydrogen (the simplest atomic species) using the newly discovered "spectroscope." Apparently hydrogen in certain circumstances did not emit light in a continuous band but in a list of discrete frequencies. Soon it was discovered that an atom's "spectrum," the special frequencies at which it emitted or absorbed light, was as characteristic of that atomic species as the shape of the beak was for Darwin's finches. (In an episode that Mach might have pondered, the discovery in the 1860s of spectral lines from known elements in the light from distant stars refuted Auguste Comte's declaration that "we shall never be able by any means to study their chemical compositions. . . . ")

More detailed properties of atomic spectrums were soon revealed. Hydrogen was found to have several series of spectral lines: one in the visible region, another in the infrared, and others still. Balmer guessed a formula in 1885: the spectral lines of hydrogen could be expressed by differences of the reciprocals of the integers, squared, such as $\frac{1}{4} - \frac{1}{9}$, or $\frac{1}{9} - \frac{1}{25}$. Spectral physics began to resemble a branch of numerology.

In 1900, the German physicist Max Planck made a proposal with even more ominous implications for the atomic modelers. Planck found a theoretical derivation of the observed "blackbody spectrum," the distribution of energy among the colors of light in a box after equilibrium conditions have been established. In the paper usually credited with initiating the quantum revolution, Planck argued that the atoms making up the walls of the cavity emit or absorb energy only in discrete chunks, or "quanta." Planck assumed that the relation between the energy and the frequency of the quantum transferred from atom to radiation was given by a simple proportionality:

$$E = h\nu,$$

energy E equals a constant (denoted by the letter h) times frequency (denoted by ν). Furthermore, the constant was universal, a new fundamental constant of nature, "Planck's constant."

Planck, an exemplary Prussian academician not attracted to radicalism, wondered whether he had simply introduced a novel accounting trick. Then Einstein (the patent clerk mentioned earlier) in March of 1905 suggested on the basis of Planck's work that light itself sometimes came in quanta, later dubbed "photons." Einstein made this startling proposal despite the successes of Maxwell's wave theory—on which he would base his own theory of relativity three months later. After Planck's and Einstein's contributions, the modeler's task seemed even more hopeless: in addition to deriving the spectral lines, the emission or absorption of energy in discrete quanta would have to be explained—and perhaps Planck's constant computed, too.

Given subsequent developments, the argument Einstein gave for his "heuristic view of the creation and conversion of light" requires an aside. Einstein noted that Planck's equilibrium formula had curious consequences. If the light in the box had long wavelengths and high density, Planck's formula said it would behave like waves trapped between the box's walls—just as Maxwell had held. But, if the wavelengths were short and the density low, the same formula predicted the light would behave as would a cloud of tiny corpuscles—as in Newton's original picture. Later in the paper, Einstein observed that his corpuscular hypothesis implied a new formula for how ultraviolet light knocks electrons out of metals, a process known as the "photoelectric effect," which was soon confirmed experimentally. Nevertheless, most physicists sensibly rejected Einstein's appeal to dual and contradictory pictures of light propagation, and "light quanta" remained controversial for decades.

In 1910, Ernest Rutherford, a New Zealander working at Manchester in England, combined a spectacular experiment with a workmanlike piece of theoretical analysis and showed that the positive charge in the atom was concentrated in a dense central nucleus. Rutherford had his assistants Hans Geiger and Ernest Marsden measure the deflection of "alpha particles," heavy, positively charged ions emitted by radium, by a thin gold foil. Some of the alpha particles were scattered back in the direction of the source. Rutherford's reaction has become the staple of all popular accounts of the birth of nuclear physics:

> It was quite the most incredible event that happened to me in my life. It was almost as incredible as if you fired a 15-inch shell at a piece of tissue paper and it came back and hit you!

The "tissue paper" Rutherford referred to was the background of diffuse positive charge in the "minnows in a pond" (also known as the "plum-pudding") model of the atom advocated by Thomson. Rutherford's explanation was that occasionally an alpha particle, aimed by chance in just the right direction, made a direct hit on a dense central nucleus and rebounded, like a tennis ball off one of the net posts. Using Coulomb's law of repulsion and Newton's laws of motion, Rutherford computed the likelihood of the alpha particle being scattered by a given angle—and his formula fit the data. The solar-system model was a success: it had explained one experiment.

In the second decade of the twentieth century, the situation in atomic physics became more and more ridiculous. The hydrogen atom, *pace* Rutherford, was supposed to be a little solar system, with the negatively charged, relatively light electron revolving about the positively charged, much heavier "proton" (a word Rutherford coined in 1920). But the electron must then radiate energy, spiraling into the proton in less time than an eye-blink. Even ignoring this catastrophe, planets in a solar system can revolve in any of a continuum of orbits, while electrons apparently had special preferences. Worse, for known emission processes—radio waves generated by currents in an antenna, as demonstrated by Heinrich Hertz in 1888, or sound waves by a tuning fork—the wave produced oscillated at the same frequency as did the source. The atom violated this rule, too. Wolfgang Pauli summed up the situation by remarking that he wished he had become a cabaret entertainer or a gambler rather than a physicist.

A natural bifurcation now occurred among theorists. Conservatives such as J. J. Thomson continued to search for the right gimmick from the standard bag of tricks that would yield a working model of the atom. One could change the force law acting between the electron and the proton from the inverse-square law of Coloumb to some more complicated law. One could add more particles. Constraints as for a rigid body might be assumed. And so forth. With so many possibilities, why despair? But models fabricated from conventional materials had a depressing habit of working only up to the date of publication.

And then there were the "progressives" ("radicals" is too strong a word), followers of Planck and Einstein, ready to entertain fresh hypotheses and not worry too much about logical or physical coherence at the outset. In 1913, a 27-year-old Danish physicist, Niels Bohr (Ph.D. from the University of Copenhagen, 1911), constructed a theory of atoms that lent some order to the chaos.

Bohr's model was elegant, if implausible. Bohr took Rutherford's solar system and simply legislated flight paths for the planets. He postulated that the angular momentum of the electron in its orbit around the proton can take only one of a discrete series of values. This is as natural as insisting that all skaters at the ice show twirl at two revolutions per second, four revolutions per second, and so on, with all other speeds forbidden. But it worked: Bohr obtained a sequence of allowed orbits which he could identify with the "stable states" of the hydrogen atom. He then made the infamous proposal that during emission or absorption of light, the electron "jumps" from one allowed orbit to another. For the frequency of the light he invoked Planck's formula, setting it equal to the energy *difference* between the two orbits, divided by Planck's constant. From these hypotheses Bohr derived Balmer's mysterious relation, even getting the right numeric factor in front (Rydberg's constant).

Bohr's scientific method at the time might best be described as opportunistic: he simply lifted the relevant formulas from where they were available. For the dynamics of the electron's orbit, he used Newtonian mechanics. For the description of the light emitted, he assumed Maxwellian theory. But neither theory provided any justification for restricting orbits to discrete series or any motivation for an electron to "jump" from one orbit to another. Bohr's model was a chimera: a quantum head grafted onto a classical body, with a tissue of ad hoc assumptions holding them together.

Although Bohr's model is sometimes lauded in textbooks, Bohr's contemporaries found it grotesque. Otto Stern and Max von Laue, two young physicists, hiked up a mountain after hearing Bohr lecture and swore that if Bohr turned out to be right they would leave physics. Bohr, of course, was aware of the dubious model of scientific reasoning his theory represented, and he went about asking everyone: "Do you believe it?" Later in life Bohr liked to tell the story of the visitor to a Danish farm who notices a horseshoe hanging over the stable door. The visitor asks the owner whether he believes it brings good luck, to which the owner replies: "Of course not. But I am told that it works even if you don't believe in it." Eventually Bohr became the leader of those theorists dissatisfied with the status quo. When Sommerfeld produced a cleaner and more coherent version of his theory (relabeled "the old quantum theory" after 1925), Bohr didn't like it. He hoped for a truly radical solution.

This episode raises many questions about the progress of science, which unfortunately lie outside the main focus of this book. (They have been discussed to a certain extent in the literature; see the Notes at the back of the book.) To my mind the most interesting are the following. What special attributes of personality or philosophical outlook are required for an otherwise respectable scientist to put forth a theory utterly lacking a rational foundation? Others had noted Bohr's orbit condition—Einstein once remarked that he had similar ideas but did not have the pluck to develop it—but only Bohr seized on it. Is the act the prerogative of youth? Einstein was 27 when he proposed the photon hypothesis, arguably the most extravagant claim of his career; Bohr was the same age when he proposed his atomic model. Finally, are temporary and expedient theories a part of normal science, or does the elaboration of such chimerical theories indicate a "prerevolutionary" condition? (The recent "fifth-force" and "cold fusion" episodes make it clear that an outbreak of wild hypothesizing does not a revolution make. Perhaps the existence of dual houses having little or nothing in common—classical mechanics and Maxwell's theory around 1900, for instance, or quantum mechanics and general relativity today—is more indicative of such periods.)

Many other experiments contributed to physicists' growing sense of desperation in those years, but one in particular stands out as the most exemplary of quantum phenomena: the Stern-Gerlach experiment.

In 1922, Otto Stern and Walther Gerlach passed a beam of silver atoms through an inhomogeneous magnetic field and noted that the beam split in two (see Figure 1; O = oven; EM = electromagnet). The deflection of the atoms by the magnetic field was not itself surprising, since a toy magnet will experience a net force in this kind of field. (The field tugs more on one pole of the magnet than it pushes on the other.) What surprised Stern and Gerlach was that the beam diverged into separate and distinct beams. If the atoms were really little magnets, with a north pole at one end and a south pole at the other, the amount of deflection should have been determined by the atom's orientation with respect to the magnetic field. If the source of the beam did not favor one orientation over another, the atoms' orientations being determined essentially by chance, the particles should have fanned out over a band between the poles of the magnet. But this is not what was observed.

Physicists believed then that an atom's magnetism is due to the whirling of the electrons around the nucleus, and possibly also to the electron spinning around an axis, like a top. The Stern-Gerlach experiment showed that this spinning motion was always "quantized," coming in fixed mul-

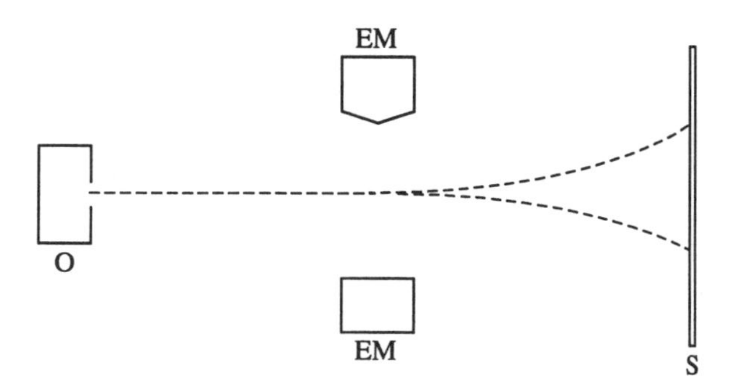

Figure 1

tiples of a fundamental unit rather than varying continuously. Moreover, the atoms in the experiment behaved the same way no matter what angle Stern and Gerlach chose for the orientation of their electromagnet. Always the initial beam split into two beams (or a whole number greater than two, depending on the atomic species used) along the chosen direction.

Confronted with Bohr orbits, jumping electrons, and the inexplicable discreteness of atomic magnetism, it was only natural that some physicists in the 1920s concluded that classical ideas were impotent to explain atoms. But where could new concepts be found?

3.

Revolution, Part I: Heisenberg's Matrices

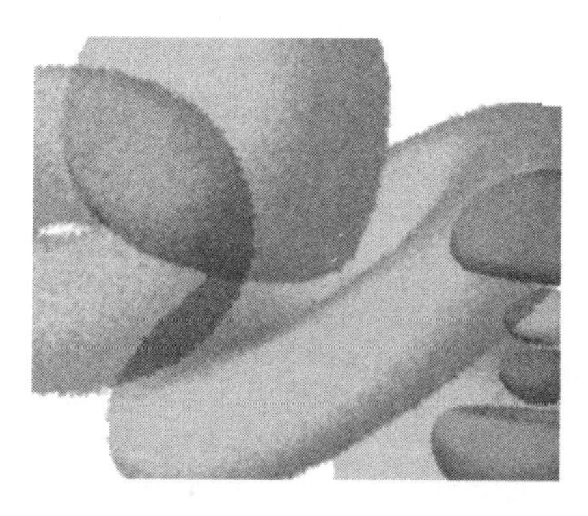

Werner Heisenberg, co-discoverer of quantum mechanics, was born on December 5, 1901, in Würzburg, Germany, the second son in a scholarly family. His father, August Heisenberg, initially taught at a gymnasium but after his *Habilitation* was appointed Professor of Greek Philology at the University of Munich in 1911. Werner's teachers in Munich made observations about the 10-year-old which seem prescient today (but which probably characterize bright children everywhere): "He has an eye for what is essential and never gets lost in details... great interest and thoroughness, and ambition." Heisenberg excelled over the other boys at skiing and chess.

At 16 or 17 (he could not recall), Heisenberg taught himself calculus in order to tutor a friend of the family, a 24-year-old chemist who wanted to get her doctorate. This showed diligence but, except for the part about the chemist, was not particularly surprising. (Today calculus is routinely taught in high school.) Perhaps more precocious was Heisenberg's early interest in philosophy: he carried a copy of Kant's *Critique of Pure Reason* with him to a farm during his mandatory war service but quickly discovered that farm labor left little time for contemplating categories of the understanding. Classical philosophy was also an important early influence; Heisenberg recalled learning about atomism "from its source" in Plato's *Timaeus* dialogue. Hiking across Germany in the company of the Wandervögel, the "free youth," was Heisenberg's other joy in the unsettled period after the disastrous Great War.

Heisenberg's diligence won him a scholarship in 1920, the examiners remarking on his "splendid and unusual talent for mathematics and physics. He was able to show... that his independent studies have taken him far beyond the demands of school." He began his studies at Munich University, and immediately one of those lunatic events occurred that make life so much like a pinball game. Heisenberg consulted the mathematician Ferdinand Lindemann—who had secured his place in mathematical history by proving in 1882 that π is transcendental—about his future. (A transcendental number is one that cannot be obtained as a root of any polynomial equation with integer coefficients. Lindemann's theorem meant that no finite series of arithmetic operations, including taking arbitrary roots, would ever turn an integer or fraction into π. Thus the simplest figure to draw—the circle—has a perimeter that transcends arithmetic.) Lindemann asked him what he had read, and Heisenberg mentioned Hermann Weyl's *Space, Time, Matter*. Weyl's book was not an easy read even for a seasoned physicist, and I suspect that Heisenberg was trying to impress the old man. (In a precollege interview I once claimed to have read all three volumes of Feynman's *Lectures in Physics*.) Whether Heisenberg was being truthful or not, Lindemann told the young man that he was "already spoilt" for pure mathematics and should therefore pursue another career. So, after consulting his father, Heisenberg enlisted in a course of studies with the theoretical physicist Arnold Sommerfeld.

Lindemann's ridiculous advice set Heisenberg on the ideal path for an ambitious young scientist in the 1920s. Atomic physics was in a wonderfully confused state, with its only viable theory of obviously transient importance and a storehouse of experiments waiting to be explained. Moreover, Sommerfeld was widely considered to be the best teacher of theoretical physics in Europe. (Einstein to Sommerfeld: "What I really admire about you is that you have produced so many talented young scientists.... You must have a gift for enriching and activating the intellects of your audience.") Scientifically, Sommerfeld was known for his development of Bohr's model of the atom, which by the 1920s was sometimes called "the Bohr-Sommerfeld model." Heisenberg's fellow students in Munich would be of the quality of his new friend Wolfgang Pauli, who, although only a year and a half older, had already published a monograph on relativity theory. Pauli became Heisenberg's friendly critic and sounding board. "I don't know how often he told me, 'You're stupid' and suchlike. That helped me a lot," Heisenberg later recalled.

Sommerfeld immediately confronted the 20-year-old with a problem at the forefront of atomic physics—the "anomalous Zeeman effect," the then-inexplicable splitting of spectral lines of an atom in a magnetic field—which Heisenberg was supposed to tackle independently. (Contrast this with contemporary university teaching methods. Who today would dare give a first-year graduate student an unsolved problem to wrestle with?) Heisenberg came back a few weeks later with a numerical account of the peculiar lines (although not a theoretical explanation), in which the young researcher violated established rules by using half-integral quantum numbers. Incredibly—recall that Heisenberg was still in his first semester—he persuaded the world's leading expert on atomic spectra that he was right.

Two more episodes must be described before relating the events of 1925. The first proved crucial to the formation of Heisenberg's philosophy of science and is among the most important in this story. The second is the quintessential "Heisenberg anecdote" and so must be repeated, however banal.

In June 1922, physicists interested in atomic theory from all over Europe gathered in Göttingen for a "Bohr-Festspiele." Heisenberg first encountered Bohr there, and he promptly got into an argument over some point in atomic theory. (Other physicists recalled staring at the blonde young man from Munich in astonishment.) Bohr later invited him for a walk, and the two discussed physics and philosophy. Thirty years later, Heisenberg recalled being strongly impressed with Bohr's epistemological views:

> Thus I understood: knowledge of Nature was primarily obtained in this way [by an intense preoccupation with the actual phenomena], and only as the next step can one succeed in fixing one's knowledge in mathematical form and subjecting it to complete rational analysis. Bohr was primarily a philosopher, not a physicist, but he understood that natural philosophy in our day and age carries weight only if *its every detail* can be subjected to the inexorable test of experiment.

(My emphasis.) Bohr later remarked to friends, "He understands everything," and invited Heisenberg to visit him in Copenhagen.

The anecdote concerns Heisenberg's doctoral exam at Munich (July 1923) and has probably been repeated at every physics thesis defense since then. It inevitably produces a chuckle. The joke can only be understood in light of later events (it relates to Heisenberg's microscope *Gedanken* experiment of 1927; see Chapter 5). Asked about his exam many years later by historian Thomas Kuhn, Heisenberg said ("Wien" was the famous experimentalist Wilhelm Wien, who had grown testy about all the talk of "geniuses" in Sommerfeld's seminar):

> I don't mind talking about that. The only danger is that I've told the story too often and improved on it every time.... Wien asked me... about the Fabry-Perot interferometer's resolving power [Heisenberg was supposed to have prepared for his exam by carrying out measurements with it]... and I'd never studied that. Of course, I tried to say something during the examination but didn't succeed.... So [Wien] certainly saw that I simply hadn't been interested. Then he got annoyed and asked about a microscope's resolving power. I didn't know that. He asked me about a telescope's resolving power, and I didn't know that either.... So he asked me how a lead storage battery operates and I didn't know that.... I am not sure whether he wanted to fail me. There was probably a fierce discussion with Sommerfeld afterwards.

Heisenberg did not fail but received a "III," the lowest passing grade in physics.

Next Heisenberg moved to Göttingen, which then had the strongest mathematics faculty, including Felix Klein and David Hilbert, in the world. Heisenberg had come to work with a fortyish theoretical physicist, Max Born, who was known for his studies of the specific heat of crystals. Heisenberg found Born too much on the mathematical side for his liking and frequently complained about the Göttingen contingent's love of formalism to Pauli. At that time it was generally agreed in Göttingen and Copenhagen that a new mechanics of particles—a "quantum mechanics"—should be the goal of research. Heisenberg began to work directly with the spectral lines of atoms, trying to guess the outline of this new mechanics from their relationships. Working with directly measurable quantities rather than more conventional variables (such as electron positions or velocities) had become a matter of principle with him, so much so that the abstract of his first paper on quantum mechanics reads, in its entirety:

> The present paper seeks to establish a basis for theoretical quantum mechanics founded exclusively upon relationships between quantities which are in principle observable.

Thus Mach's philosophy was still influential a decade after his death. Curiously, Heisenberg could not later recall having read Mach—which is understandable for one so young (23). He expressed the opinion that his positivism came from studying Einstein's relativity theory, an important clue to be discussed later.

Now we come to the least understood events of 20th-century physics. It all started with an attack of hay fever.

After a semester-long visit in Copenhagen in 1924–25 ("The past six months were the best time in my studies to date as far as science is concerned," he wrote to Bohr in April), Heisenberg spent the spring of 1925 in Göttingen. He struggled to find the long-sought quantum mechanics, but to no avail. On the seventh of June, he suffered an allergic attack. "I couldn't see from my eyes," he recalled, and "was just in a terrible state." He left on the night train for the island of Helgoland in the North Sea, where grasses did not grow. Upon his arrival the next morning, his landlady remarked that with his swollen face he must have been beaten up on the train. In a flush of improved health, Heisenberg began working strenuously on the quantum problem, pausing only to clamber over the boulders with a book of poetry in hand. One night he had it. "It was about three o'clock when the final result of the calculation lay before me.... At first I was deeply shaken. I was so excited that I could not think of sleep. So I left the house... and awaited the sunrise on the top of a rock."

Heisenberg did not solve the hydrogen atom that night in Helgoland. He actually solved a "toy problem"—a simple oscillator—but that is the usual first step in theoretical physics. What lead to his rapture on the rock that night was the appearance of quantum phenomena in one concrete problem.

Heisenberg wrote his first paper in July and sent it to *Zeitschrift für Physik.* It is very difficult to understand, as might be guessed for the most revolutionary work of the century, and especially one written by so young an author. There are startling leaps in logic, baffling appeals to analogies between classical and quantum, and bizarre mathematical objects casually mixed with familiar formulas. In short, it is a fascinating work.

Heisenberg begins by objecting to electron orbits as being unobservable. So, he says, one must introduce a new kinematics prior to introducing a new mechanics. The reader grasps that the physical description of particle motion, even in the absence of forces, will suffer some alteration, and Heisenberg immediately fulfills this expectation with a transformation that must have sent most physicists reeling. The position of the electron enters into classical theory, Heisenberg notes, either as a quantity squared or with a higher power. But if the position as a function of time, that is, the *orbit*, will lose its classical meaning in quantum mechanics, how (asks Heisenberg) will one express the square of this quantity? (Already I am lost. Can the square of a quantity somehow retain meaning if the quantity itself is bereft of it?) Heisenberg next writes out a classical expression for the orbit in terms of Fourier series (named after the 19th-century mathematician and physicist Joseph Fourier), a means of decomposing motion into a sum of simple oscillations, each of a definite amplitude and frequency. (The familiar decomposition of the complicated motion of a guitar string into a fundamental mode, a first overtone, etc., is one application of Fourier series.) He remarks that in classical theory one can easily compute the square of the position in terms of the Fourier amplitudes and frequencies.

Now comes the leap beyond logic that accompanies every genuinely revolutionary work. Heisenberg declares that in quantum theory one should replace the Fourier amplitudes by certain square arrays of numbers constructed from the spectral lines observed in the laboratory. With this virtual non sequitur, Heisenberg crossed over from the "classical" era to the modern one. The period in which one could visualize events in the interior of the atom, which had lasted a mere dozen years, was at an end.

On the interpretation of what one has got after this strange replacement, Heisenberg is silent. He proceeds nevertheless to compute the square of the position as promised, introducing as he goes a rule for "multiplying" two square arrays in such a way as to get another one. In this step Heisenberg rediscovered the law for multiplying matrices—square or rectangular arrays of numbers—which had been known to mathematicians for decades but was not part of the physicists' standard toolkit at the time. (Today college or high-school science students routinely study n-by-n matrices, where n might be three or four; Heisenberg's matrices, by contrast, were "infinity-by-infinity," and even today these are usually reserved for graduate school.) With the breezy assertion that "it is very natural to take over the equation of motion directly into quantum theory," Heisenberg plugs his infinite arrays into the classical equations. (This step invariably reminds mathematics teachers of the beginning calculus student's desire to plug the symbol for "infinity," the infamous eight-on-its-side, into every formula encountered. It usually takes an entire semester to stamp out this regrettable tendency!)

Having taken these frightening steps, Heisenberg retreats to more conventional lines of thought. Bohr's quantum theory was based on a law of motion (the classical one) and a quantization condition (the ad hoc restriction on the electron's angular momentum). Heisenberg suggests that one should try to do the same with his square arrays. This was an attack of uncharacteristic timidity: Bohr's quantization condition must *follow* from his assumptions to lend them credibility. He notes that the hydrogen atom seems too complicated for his methods. And he worries that perhaps he has inadvertently nullified the law of conservation of energy. But, putting aside these qualms, he demonstrates that some quantum relations concerning emission and absorption of light quanta follow from his hypotheses. He ends this first paper of modern quantum theory by treating two other cases, an anharmonic oscillator and a rotor.

There was much reasoning by analogy in Heisenberg's first effort. This was typical of papers in quantum theory at the time, partly because no one knew the "right" equations yet, and partly due to another of Bohr's dictums. Bohr knew that any quantum theory would have to "save the phenomena" of the macroscopic world. Cannon balls and planets are correctly described by classical physics; hence quantum mechanics must approximate classical mechanics in certain circumstances. Bohr proposed that the two theories should come into correspondence in the limit of "large quantum numbers" (essentially orbits very far from the nucleus), which became known as Bohr's "correspondence principle." (Later Bohr elevated classical physics even higher, claiming that it can *never* be abolished or absorbed by another theory; see Chapter 6.) This was the other principle guiding Heisenberg's thinking that summer, in addition to his "observability requirement"—also acquired, at least in part, from Bohr.

Back in Göttingen, Heisenberg gave his paper to Max Born, who was immediately fascinated and moreover equipped with a crucial piece of information:

> After having sent Heisenberg's paper . . . for publication, I began to ponder about his symbolic multiplication, and was soon so involved in it that I thought the whole day and could hardly sleep at night. For I felt that there was something fundamental behind it. . . . And one morning . . . I suddenly saw light: Heisenberg's symbolic multiplication was nothing but the matrix calculus, well known to me since my student days. . . .

Born settled down to work with his student Pascual Jordan (age 22), and in a few months they eliminated several of Heisenberg's worries: they proved that both Bohr's quantization condition and the law of conservation of energy followed from Heisenberg's assumptions. They based their work on perhaps the strangest equation written down in the four millennia since the Babylonians invented multiplication:

$$\mathbf{QP} - \mathbf{PQ} = \frac{\sqrt{-1}\,h}{2\pi}.$$

In this equation **Q** and **P** denote two of Heisenberg's infinite matrices, which "stand for" the position and momentum of a particle, respectively, whatever this means, and h is Planck's constant. (Clarification would come later.) The left side is written so as to emphasize a dramatic feature of matrix multiplication: it is not "commutative." That is, order may make a difference. For matrices, **P** times **Q** does not have to equal **Q** times **P**, as is the case for ordinary numbers. On the right side of the Born-Jordan equation, we have not zero but the square root of minus one as well as some constants, one physical and one mathematical. Thus, the equation states that **P** and **Q** do *not* commute, the measure of the discrepancy being $\sqrt{-1}$ times these constants. Heisenberg recalled being "terribly worried" at first by the noncommutativity of his number arrays.

Complex numbers—numbers with real and imaginary parts—had been used extensively as a tool by scientists since Cauchy and Gauss popularized them in the previous century. But this was the first use of the imaginary unit *in a seemingly essential way* in the history of natural science. (A year later Schrödinger would discover to his great surprise that his wave had to be assigned a complex "height.") It was also the first time that two mathematical entities representing physical quantities had been shown not to commute when multiplied. Thus, two fanciful possibilities from the dreams of mathematicians had entered into physics with a bang. (As indications of the importance physicists attach to the Born-Jordan equation, one notes that it was incorporated into a stained-glass window in Old Fine Hall at Princeton, and it appears on Born's tombstone.)

Events moved rapidly after these discoveries. In England, Paul Dirac (age 23) showed that the new mathematics could be interpreted as a "Hamiltonian dynamics of matrices"—an essentially classical dynamics for Heisenberg's arrays. (Dirac wrote the first Ph.D. thesis on quantum mechanics a year later.) Heisenberg, Born, and Jordan collaborated on a long paper on matrix mechanics which, among other things, made it clear that the mathematicians had paved the way for the physicists: much of the algebra they needed had already been developed by David Hilbert. As a result, quantum mechanics is said today to be a theory set in "Hilbert space." Then Wolfgang Pauli (age 25) claimed victory for the new theory by computing the spectrum of the hydrogen atom from matrix mechanics, without additional assumptions.

The reader may have noted that these great successes were attained partly by men barely out of childhood. Indeed, this became something of a joke in Göttingen, where people began to refer to *Knabenphysik* (kid physics). But I suspect that the older scientists, such as Niels Bohr and Max Born, had as important a role to play as the youngsters in these developments. In collaborations among scientists at various stages of their careers, it is generally the older scientists who choose the direction of research and legitimize or repudiate the ideas of their younger colleagues. Sometimes, of course, their conservatism stifles the younger scientists' creativity—but neither Bohr nor Born was conservative. The role of the youngsters, on the other hand, with their boundless enthusiasms, unformed epistemologies, and original techniques, is to break through the logjam in a blocked research program. It is for this reason that thoughtful older scientists like to work with younger men or women in their twenties or early thirties. Of course, it helps if they are geniuses, or at least very bright.

Thus far, particles had dominated the thinking of the quantum revolutionaries—although the classical properties of particles were, like Lewis Carroll's Cheshire cat, disappearing bit by bit, and perhaps only the grin would remain behind. But before the meaning of what they had done could be made plain, a rival quantum theory appeared that made no mention of particles.

4.

Revolution, Part II: Schrödinger's Waves

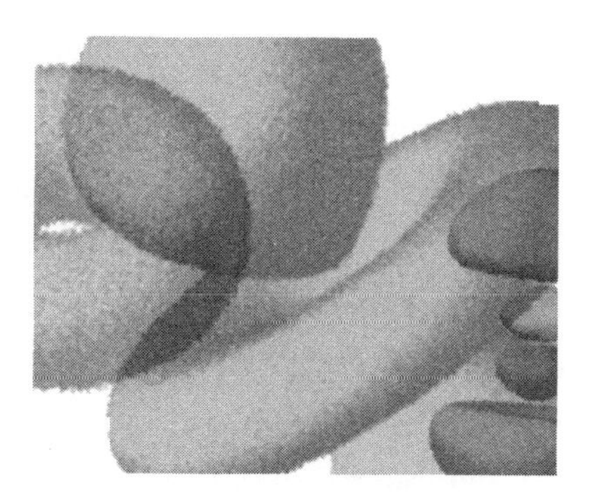

Erwin Schrödinger, co-discoverer of quantum mechanics, was born to middle-class, intellectual parents in Vienna in 1887. His father, who owned a linoleum factory, was an enthusiastic amateur botanist. His grandfather on his mother's side was a chemist and professor at the Polytechnicum (the technical university) in Vienna. Erwin, an only child surrounded by adoring adult women (mostly maiden aunts), had a happy upbringing. He showed early talent at mathematics and was the most promising pupil at his Gymnasium.

In the fall of 1906, Schrödinger enrolled in the University of Vienna, intending to pursue a curriculum in physics. He had hoped to study gas theory under the great Ludwig Boltzmann, but Boltzmann's suicide prevented this. Instead, his teacher would be Boltzmann's student and successor, the young, enthusiastic theorist Fritz Hasenöhrl, who had nearly attained immortality by writing down $E = Mc^2$ before Einstein. (Unfortunately, he got the c^2 factor wrong.) Mach, although retired after his stroke, was still active—so the philosophical tides that would batter the mature scientist ebbed and flowed around the young scholar, who noted the opposed schools but with youthful optimism dreamed of their eventual reconciliation.

Schrödinger was the top student of physics and mathematics in Vienna. Fellow students sometimes pointed to him with awe: *Das ist der Schrödinger*, they would whisper. This initial promise, however, was not immediately realized. Schrödinger completed his doctorate (roughly the equivalent of the American master's degree) in 1910 with an unexceptional treatise on the conduction of electricity in moist air. The formal *Habilitationschrift*, permitting him to apply for a junior position (*Privatdozent*) at the university, came two years later. Combining several works on electricity and magnetism, it was also not awe inspiring; although the mathematics were brilliant, the physical assumptions were often wrong—a scenario that would haunt Schrödinger in subsequent years.

Schrödinger had unusually broad interests for a scientist. At one point he became the world's leading expert on *color theory*, the science of the perception of colors. He studied philosophy, especially Eastern mysticism, and wrote romantic poetry. At times he considered his philosophical writings or his poems to be on a par with his physics. (Few philosophers or poets agreed.) The pursuit of women, especially young women, was his other major preoccupation. Later in life, after fleeing the Nazis, Schrödinger arrived in Catholic Ireland accompanied by a wife and mistress; but it was only after he became infatuated with a colleague's daughter that discrete intervention was deemed necessary.

In the Great War, Schrödinger served as an artillery officer on the Italian front. His teacher Hasenöhrl was killed, as were so many other young scientists, poets, and artists. After the war Schrödinger pursued a successful if unspectacular career as a theoretical physicist, ascending the academic ladder with positions in Jena, Stuttgart, Breslau, and finally Zürich. During this time he wrote many papers in statistical mechanics and made minor contributions to atomic theory and radioactivity.

In the winter of 1926, Schrödinger was 38, married, secure in his position, but without world-class attainments. Then, in what his mathematician friend Hermann Weyl called a "late erotic outburst," Schrödinger took a recent doctoral thesis and his current mistress up to a villa in the Swiss Alps and set out to make his scientific fortune. (I mention these facts in case the reader would like to further pursue Weyl's erotic theory of scientific creativity. While old stuff where artists or poets are concerned, for scientists this might be fresh ground; Ph.D. candidates, take note.) He was going to construct a new theory of matter.

The doctoral thesis dated from two years before and was written by a young French physicist named Louis de Broglie. (Of the mistress nothing is known, not even her name.) De Broglie (born 1892, then a student of Paul Langevin at Paris) speculated that types of matter other than radiant energy, including electrons, might have wave as well as particle properties. This was a proposal similar in spirit to Einstein's introduction of the light quantum in 1905, but coming from the reverse side: no one doubted the electron's qualifications to be a particle, a fact demonstrated to everyone's satisfaction years before. De Broglie was suggesting that electrons might also, in some circumstances, display wave properties such as diffraction. He extended Planck's mysterious formula relating energy to frequency to electrons, just as Einstein had to light quanta, and derived an additional relation:

$$p = h/\lambda;$$

momentum p equals Planck's constant h divided by wavelength λ. By connecting momentum and energy (point-mechanical quantities) with frequency and wavelength (properties of a continuum), he completed a formal dualism between particles and waves. One could now "understand" Bohr's quantization condition for the hydrogen atom, de Broglie argued, as the requirement that an integral number of wavelengths fit exactly along each circular orbit.

Impressed by de Broglie's suggestion (and some remarks of Einstein along similar lines), Schrödinger conjectured that matter might be built entirely out of waves. The stable states of the hydrogen atom might be explained by a fundamental vibration, a first overtone, and so on—just as for a vibrating drumhead. And for a free electron, which was capable of leaving such suggestive tracks in cloud chambers, Schrödinger imagined a little packet of waves moving along without dispersing. In exploring this possibility, Schrödinger had another resource to spur him on (in addition to the view of Alpine splendor out the window, and the charms of the young woman in his bed). That was his friendship with the aforementioned mathematician, Hermann Weyl.

Weyl was expert in many topics in mathematical physics, including the non-Euclidean geometry exploited by Einstein a decade before to construct a new theory of gravity. (In *Space, Time, Matter* (1918), the book young Heisenberg claimed to have read, Weyl had advanced his own version of relativity, including an attempt to unify gravity with electromagnetism that had drawn criticism from Einstein.) Most importantly from Schrödinger's point of view, Weyl knew the theory of "proper vibrations" in continuous media. This well-developed mathematical discipline treated standing waves in a variety of situations, including water in a lake, air in an organ pipe, and strings on a guitar. Schrödinger hoped it would apply to the atom as well. With Weyl's help, he succeeded in deriving Bohr's stable states of the hydrogen atom, without any recourse to the classical picture of the electron as a point particle.

The professor, long past his prime according to the established myths about scientific creativity, had succeeded in his first goal: he had derived the structure of one kind of atom from entirely new principles. Schrödinger sent his first paper on what soon became known as "wave mechanics" off to a journal, and told his wife he was going to win a Nobel Prize.

But what were these new principles? Schrödinger was far from sure what he was doing in this first effort. His starting point was Hamilton-Jacobi theory—a 19th-century approach to solving Newton's problem of particle motion with a geometric flavor. William Hamilton's and Carl Jacobi's idea was to find a family of surfaces in space such that the curves perpendicular to the surfaces are mechanically possible motions. One can think of these curves as paths of a cloud of imaginary particles, all moving according to Newton's laws without interacting or colliding; a real particle can then be singled out by specifying its initial conditions. Since the surfaces in Hamilton-Jacobi theory can be thought of as wave fronts, this approach tries to solve a problem in mechanics as though it were a problem in wave propagation.

Given de Broglie's insight, the Hamilton-Jacobi theory was certainly a plausible starting point. However, with a few transitional remarks Schrödinger whisks away the classical recipe and replaces it with another, a so-called variational problem. He plugs a wave into the classical expression for the energy of a particle and asks that the resulting quantity, integrated (totaled) over all of space, be at an extreme value—a high or low point of "total energy" relative to a variation of the wave configuration. Now, a variational principle as the basis of a dynamical theory was not a new idea. The "principle of least time" described how a light ray moved through a medium, and Hamilton had shown that all of mechanics could be based on a "principle of least action." But a principle of least energy? It sounded more like a description of work ethics in a government shop than a new principle of atomic physics. This variational principle was original—and baffling.

A few months later, Schrödinger tried to explain the physical reasoning behind his mathematics. In his second paper he admits that the steps in his transformation of Hamilton's problem into his own were "incomprehensible." But, Schrödinger points out, the wave in Hamilton-Jacobi theory included only the *geometrical optics* limit of a true wave theory—a useful approximation in which the wave is assumed to propagate in straight lines (curves in an inhomogeneous medium), neglecting all effects due to diffraction. (It is diffraction that allows light to appear in a region behind a small obstacle such as a strand of hair, which would be completely in shadow if light moved rectilinearly.) But de Broglie had proposed that electrons have true wave properties. Therefore, says Schrödinger, when dealing with situations such as the hydrogen atom in which the electron's orbit must bend sharply around the proton, the geometrical optics limit must be replaced by a true wave description.

It was beginning to sound plausible—although that last remark reminds one of a magician applauding his assistant after making her vanish. Certainly Schrödinger's derivation of the hydrogen spectrum in a continuum theory was a mathematical triumph. Were "point particles" destined to go the way of phlogiston and the aether?

By mid-winter 1926, the crisis was apparent. There now existed *two* new quantum theories, both radical departures from classical physics. They seemed as irreconcilable as... well, waves and particles.To make matters worse, no plausible physical interpretation was available for either one. In Schrödinger's theory, was the electron actually spread out over the whole volume of the hydrogen atom—or even through all of space? In Heisenberg's, how could a square array of numbers "represent" a particle? And what did the strange "non-commutitivity" of position and momentum mean?

Schrödinger was the first to shed some light on the mystery. The third paper of Schrödinger's great tetralogy on wave mechanics is titled "On the Relationship of the Heisenberg-Born-Jordan Quantum Mechanics to Mine," and in it Schrödinger proves an almost unbelievable fact: the complete mathematical equivalence of the two theories. Schrödinger's words reveal his astonishment:

> In view of the extraordinary divergence of the starting points and concepts of Heisenberg's quantum mechanics, on the one hand, and the theory recently given here its basic outlines and called "undulatory"... mechanics, on the other hand, it is very strange that... these two new quantum theories agree with one another even where they differ from the old quantum theory [i.e., Bohr's theory].... This is really very remarkable, since the starting point, concepts, method, and the whole mathematical apparatus in fact appear to be different...

Schrödinger next explains how certain calculus operations performed on his wave function yield Heisenberg's square matrices. In passing he remarks that physical equivalence is not the same as mathematical equivalence—an observation that would prove more prescient than Schrödinger probably realized. Finally, apologizing for the "one-sidedness" of his views, he argues that his approach is superior to Heisenberg's in that it does not entirely suppress intuition by operating with purely "abstract concepts" such as energy levels or quantum jumps. (I presume he meant as opposed to his more concrete vibrating wave. He need not have apologized: much nastier comments about his theory—"wretched physics," "an abomination"—were heard in Göttingen and Copenhagen.)

But Schrödinger had left one issue dangling. Since atoms occasionally make transitions from one stable state (represented in Schrödinger's theory by a proper vibration) to another, it is crucial to know how the wave evolves in time. Schrödinger begins his fourth communication on wave mechanics with this question. The conventional wave equation describing the evolution of sound or water waves will not do, he remarks, since it presumes a fixed value of the energy. There is another equation that seems preferable, but it leads to difficulties when the energy is varying. The situation can be improved, writes Schrödinger, by factoring the equation using *the square root of minus one.*

Schrödinger's next remarks reveal the agony of creativity. He requires that his wave henceforth satisfy the simpler equation obtained by factoring. But this new equation contains the imaginary unit, so the solution may contain it as well. Schrödinger tries to gloss over this frightening possibility, noting correctly but irrelevantly that one can still regard the real part of the solution as a "real wave function." (Schrödinger has forgotten that the real part by itself will not satisfy any equation and is therefore useless for making predictions.)

By an implausible route, and accompanied by unneeded baggage, Schrödinger has arrived at the Finland Station. Before this paper, physicists tried to discover the laws of motion of the particles making up atoms, with limited success. Afterwards, they will solve Schrödinger's equation. More precise predictions of natural phenomena will follow from this equation than from any other mathematical relation proposed in this century.

Before coming to Schrödinger's thoughts about what it all means, a few words about the mathematical form of his wonderful equation. First, the most striking fact—besides the appearance of the square root of minus one—is that it is linear. That is, the wave variable is never squared, cubed or raised to any power but unity. As a pleasant consequence, adding two solutions (or allowed waves) gives another one—yielding the so-called superposition principle. (Naturally, theorists working in fields such as hydrodynamics, where the equations are nonlinear and famously nasty, flocked to the new field where you could solve the equations—and the physical or philosophical difficulties be hanged!) This will have unforeseen consequences.

Second, with a slight modification, Schrödinger's equation can be given a classical, even mechanical, interpretation. The modification is to replace space, a continuum, by a discrete lattice-work of points. (This is not as radical as it sounds; if the space between the points of the lattice is made small enough—say 10^{-50} cm—we could never detect this breakdown of continuity.) After doing so, Schrödinger's equation acquires the interpretation shown in the Figure 2.

Springs connect a series of points (balls in the figure) to an imaginary "floor" and "ceiling" (serving merely to make the drawing more com-

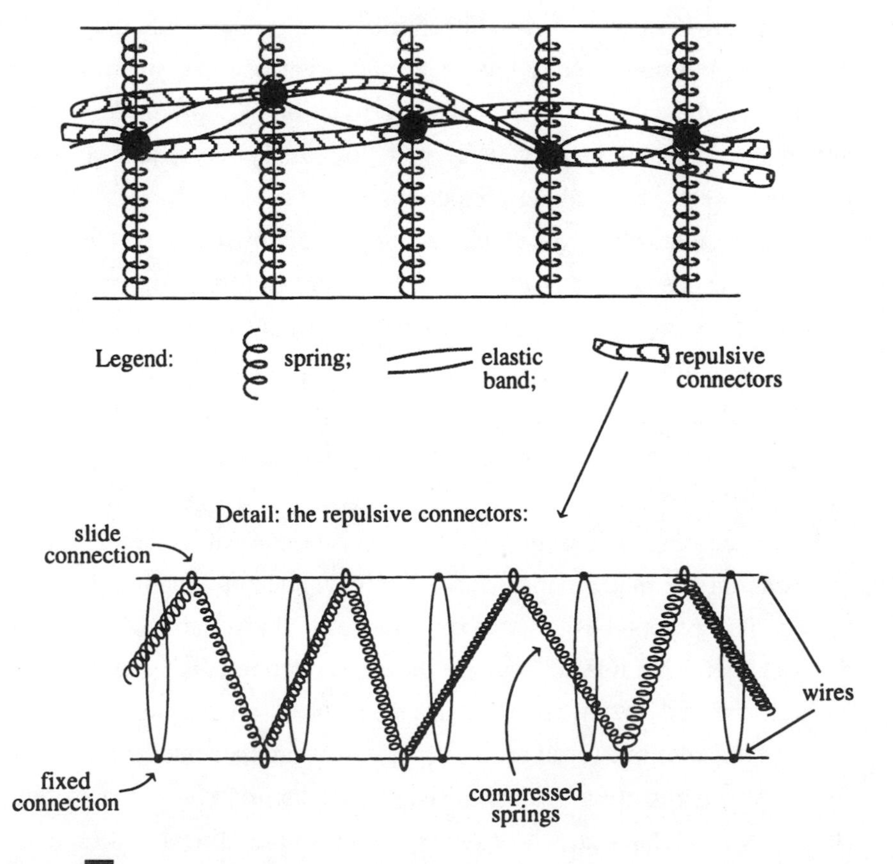

Figure 2

prehensible); also, elastic bands connect nearest-neighbor points. Slightly more complicated objects, "repulsive connectors" generating repulsive forces one-fourth as strong as the attractions generated by the bands, couple next-nearest neighbors (see the detail in Figure 2). The stiffness of the vertical springs is set by the strength of the force present (for instance, in the hydrogen atom, due to the proton's attraction of the electron); the strength of the horizontal connectors is fixed by Planck's constant. One sets the whole apparatus into motion by displacing some of the balls vertically; afterwards, it follows Newton's laws of motion. At the risk of *less-majesté*, Schrödinger's equation can be thought of as merely a clever way to compute positions and velocities in this Rube-Goldberg contraption. (It might be useful for classroom demonstrations, but I know of no attempt to actually construct it.) Unfortunately, since the whole shebang represents the state of a *single particle*, these springs and balls do not suggest a physical interpretation.

At the end of his fourth paper, Schrödinger finally turns to the physical significance of his wave function, now possibly imaginary. This is a remarkable section, as Schrödinger comes tantalizingly close to stating the probabilistic interpretation later advanced by Born. He writes:

> [My procedure] is equivalent to the following interpretation, which better shows the true significance of [the wave function]. [The square of the length of the wave variable] is a kind of weight function in the configuration space of the system... each point-mechanical configuration enters with a certain weight in the true wave-mechanical configuration.... If you like paradoxes, you can say that the system is, as it were, simultaneously in all kinematically conceivable positions, but not "equally strongly" in all of them.

What are these "weights" Schrödinger tosses out so off-handedly? The modern reader, trying to make sense of this "paradox," scans down the page for the word "probability"—and does not find it. Instead, there are these unexpected sentences:

> This re-interpretation may appear shocking at first sight after we have previously spoken in such an apparently concrete form of the "[wave]-vibrations" as of something quite real. There is a basis of some tangible reality, however, even according to the present treatment—namely, the very real... fluctuations of the electric space density.

Schrödinger still hopes his wave might represent a spread-out electron. But there is a second difficulty: for two or more particles, more than three numbers are needed to describe their locations, so his wave exists in a fictitious space of more than three dimensions. Schrödinger has not forgotten this point; he even emphasizes it a few sentences later:

> It has been stressed repeatedly that the [wave]-function itself cannot and must not in general be interpreted directly in terms of three-dimensional space however the one-electron problem leads towards this....

Unfortunately for Schrödinger, this last fact is fatal to his view.

In June of 1926, Max Born proposed an entirely different interpretation of Schrödinger's wave. (Schrödinger's mathematics caused no controversy; on the contrary, most physicists were only too happy to abandon Heisenberg's peculiar matrix algebra for partial differential equations, a familiar branch of classical mathematics.) Schrödinger had multiplied the square of his wave by e, the elementary unit of electric charge, and considered the result the charge distribution of a spread-out electron. Since the total charge does not change with time, this made sense. But leaving the factor of e out, the "something" conserved might be interpreted as a *probability*: its total over all space being one then has the simple meaning that the electron, being indestructible, is surely located somewhere at all times. And the conundrum that worried Schrödinger—his wave flowing through a fictitious higher-dimensional space in the case of two or more electrons—is simply resolved. It is perfectly natural to speak of the probability that one electron is *over here* while the other one is *over there*, and specifying the position of two electrons puts us in six-dimensional space. As Born put it somewhat later:

> It is necessary to drop completely the physical pictures of Schrödinger which aim at a revitalization of the classical continuum theory, to retain the (mathematical) formalism and to fill that with new physical content.

Ironically, Born later attributed the idea of a "ghost wave" (*Gespensterfeld*) to Einstein, who never published it. I suspect Einstein later wished he had never mentioned it.

Here are the main points in Born's interpretation of quantum mechanics, as elaborated somewhat by his colleagues:

(1) The wave is a wave of probability, not a "real" entity, like a light or sound wave. (2) The point, or at least very small, electron is "real," but its position at any time is unknown. The squared length of the wave function at each point in space gives the probability that the particle is in a small volume centered at that point. (For example, if the wave function at some point equals $3/10 + 4/10\sqrt{-1}$, its squared length equals $9/100 + 16/100 = 0.25$.) (3) Although Schrödinger's wave evolves continuously and deterministically for most of the time, during the emission or absorption of light Bohr's "quantum jumps" nevertheless occur, in which the particle makes an instantaneous leap from one "quantum state" to another. These jumps occur unpredictably, discontinuously, and are not described by Schrödinger's equation. (4) After a quantum jump Schrödinger's wave instantly "collapses" to the one representing the new state of the particle.

For at least the next six decades, physicists will debate the meaning and correctness of this peculiar amalgam of discrete with continuous, causal with random.

Born's probabilistic interpretation was as far removed from Schrödinger's original dream of a pure wave theory as the Soviet Union was from Marx's vision of a classless society. Schrödinger hated it. In conversation with Bohr and Heisenberg, he remarked that if he had known "this damned quantum jumping" was going to stay, he would never have gotten involved with quantum mechanics in the first place.

Born's interpretation was immediately adopted by Heisenberg and the other "kids" of that year of kid-physics. Most of them had already thought of it. It abandoned a causal description of events in the atom—which nicely fit certain ideas that Heisenberg was incubating. (Bohr was even more pleased, as we shall see in Chapter 6, "Complementarity.") And most importantly, it preserved the notion of point particle; in retrospect, Schrödinger's hope that particle tracks could be explained as little wave packets moving without dispersing was an illusion. Computations soon proved that even a tiny wave packet could grow to a size obviously larger than any particle track—yet the particle did not grow with it. As shown by collisions with other particles or by measuring devices such as photographic plates, it retained its "particlehood" intact. (It is unfortunate that Schrödinger did not know the mechanical analogy in the figure, which makes it obvious that any initial disturbance will quickly spread out.)

The wave idea had worked—yet the particles had refused to go away. Nature, or perhaps the Goddess of Mathematics, had played a trick on Schrödinger.

5.

Uncertainty

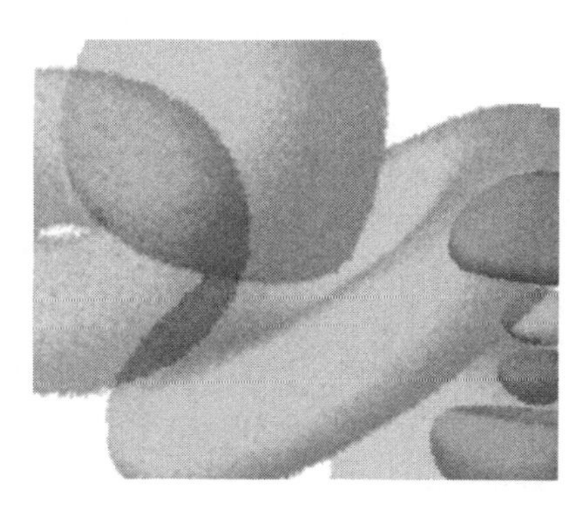

The Göttingen-Copenhagen axis had quickly sunk Schrödinger's wave interpretation, but Heisenberg was not happy. In a letter to Pauli he complained:

> Here I am in an environment that is diametrically opposed [to our view], and I don't know if I am just too stupid to understand mathematics. Göttingen is splitting into two camps. Some people, like Hilbert, talk of the great success achieved through the introduction of matrix calculus into physics, and others, like Franck [the noted experimentalist], say these matrices can never be understood. I am always angry when I hear the theory called nothing but matrix physics.... 'Matrix' is certainly one of the most stupid mathematical words in existence.

To dispel the smugness of the mathematicians and the bafflement of the physicists, Heisenberg had to discover a physical meaning for his matrices. Starting in the winter of 1927, he worked on the problem alone in Copenhagen, Bohr being away on a skiing holiday. The hardest facts to reconcile with either his matrices or Schrödinger's waves were photographs taken of cloud chambers, in which water droplets strung out like beads on an invisible string reveal the passage of an electron. Neither theory seemed capable of describing such paths. "In my despair about the futility of my attempts I remembered a discussion with Einstein and his remark: 'It is theory which describes what can be observed,' " Heisenberg recalled. He tried to turn the question around. Perhaps only that described by the mathematics of quantum mechanics can exist in nature. The water drops in the cloud chamber might serve only to delimit certain possibilities. "There was not a real path of the electron in the cloud chamber," Heisenberg concluded.

By March of 1927, Heisenberg was ready with his conclusions. The resulting "uncertainty principle" paper may be the most quoted of this century by a physicist; yet of all the great physics articles I have read (excepting only Bohr's philosophical papers), it is the most problematic. It also ends with an amazing snapper.

Heisenberg begins by remarking on the strange dichotomies that seem the essence of quantum mechanics: continuous versus discontinuous, wave versus particle. These suggest that our understanding of place and motion may be at fault. It is the first dichotomy that is the crucial one:

> When one admits that discontinuities are somehow typical of processes that take place in small regions and in small times, then a contradiction between the concepts of 'position' and 'velocity' is quite plausible.

The classical picture of a particle's path as a continuous curve might have to be replaced by a discrete series of points in spacetime, writes Heisenberg. Nevertheless,

> All concepts which can be used in classical theory . . . can also be defined exactly for the atomic processes in analogy to the classical concepts.

The breakdown of classical ideas occurs only when one attempts to measure *two* quantities, such as the position and momentum (mass times velocity), simultaneously. Then "indeterminacy" creeps in. It is this "uncertainty" that makes possible the strange noncommutativity discovered by Born and Jordan, "without requiring that the physical meanings of P or Q be changed."

As a concrete illustration of his ideas, Heisenberg introduces a *Gedanken* experiment, perhaps the best known from the early years of quantum theory: the "gamma-ray microscope" (see Figure 3). The goal in this imaginary exercise is to measure the position of a single electron

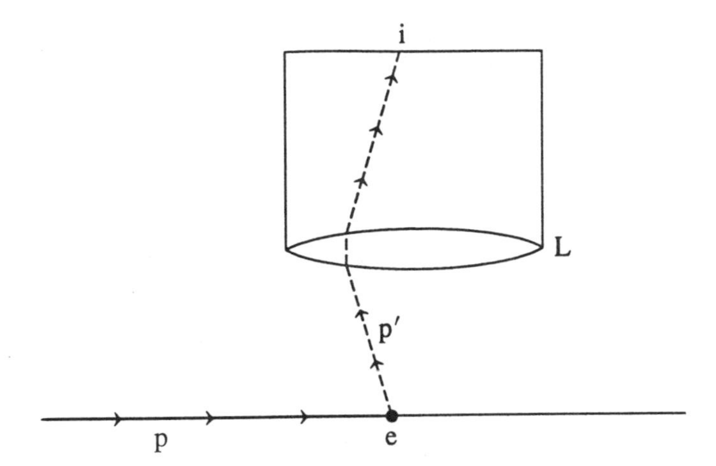

Figure 3

in empty space as accurately as possible. One might approach the task exactly as one would for a dust mote or a bacterium: by illuminating the electron and then using a microscope to focus the reflected beam into an image. Because of the electron's infinitesimal size, however, one will be forced to use "light" of extremely short wavelength. Hence the gamma rays, electromagnetic radiation with wavelengths even shorter than ultraviolet or x-rays. (Here Heisenberg leaves the laboratory for the rarified plane of the thought experiment: no one, then or now, knows how to build a microscope that uses gamma rays.)

When a gamma ray strikes an electron, Heisenberg notes, there is a discontinuous change in the electron's momentum, due to the "kick" imparted by the collision of a photon (one of Einstein's "light quantums") with the electron. Remarkably, one can treat this collision theoretically exactly as one would the collision of two billiard balls—which was fortunate for Heisenberg, as the detailed theory of such collisions would become available only much later. Applying the laws of conservation of energy and momentum, which state simply that what both particles bring into a collision they must also carry off, one can derive formulas virtually identical to the billiard case. (This fact was proven experimentally by the American physicist Arthur Compton in 1923; as the formulas are very different from those of classical wave theory, it laid to rest the lingering doubts about the reality of the photon.) Heisenberg remarks that the more precisely the position is to be determined, the shorter the wavelength and the greater the required momentum of the photon. Hence the "kick" imparted to the electron will be correspondingly greater. If δQ is the "precision" with which the position is known, and δP that for the momentum, the product of these quantities will be about equal to, but never less than, Planck's constant h:

$$(\delta P) \times (\delta Q) \approx h.$$

There is a corresponding relation for time and energy:

$$(\delta T) \times (\delta E) \approx h.$$

These are the quantitative statements of Heisenberg's celebrated uncertainty principle. Heisenberg derived the first relation mathematically, defining δP and δQ to be the statistical spread in these values and using Born's probability interpretation.

"Indeterminacy," "uncertainty," "imprecision," "statistical spread"—the terms Heisenberg uses in this paper jump around almost as much as the electrons. It is as if Heisenberg wished to illustrate his uncertainty concept semantically as well as mathematically. Scientists and philosophers have been trying ever since to discover exactly what he meant. Is it the particle that is uncertain—or is it us? If the latter, how could our ignorance cause an electron to act like a pinball? Granted that the photon perturbed the electron, why can't further measurements reveal by how much? (Such "retrodiction" is certainly possible for billiard-ball collisions.) And why should quantities like position or momentum, mere tokens of a soon-to-be-abandoned paradigm, retain precisely their old meanings in a radically new one? (Some answers that have been proposed to these questions will be discussed in Chapter 16, "Paradoxes.")

At the end of Heisenberg's paper there is a remarkable "Addition in Proof." This note commemorates a dispute between Heisenberg and Bohr in Copenhagen, which almost led to a falling out between the two men. Bohr strongly criticized Heisenberg's analysis of his gamma-ray-microscope thought experiment. Heisenberg fought back, then relented. (Heisenberg: "I remember that it ended with my breaking out in tears because I just couldn't stand the pressure from Bohr." Recall the difference in age—Bohr was 42, Heisenberg 26—and authority between these two men.) Bohr prevailed, and Heisenberg agreed to take back the nub of his thesis *at the moment of its publication*, writing:

> ...Bohr has brought to my attention that I have overlooked essential points in the course of the discussions in this paper. Above all, the uncertainty in our observation does not arise exclusively from the occurrence of discontinuities but is tied directly to the demand that we ascribe equal validity to the quite different experiments which show up in the corpuscular theory on the one hand, and in the wave theory on the other hand.

Bohr had convinced Heisenberg that it is not discontinuity but conceptual (or even linguistic) *duality* that begets uncertainty. Indeed, going over Heisenberg's *Gedanken* experiment more carefully, as we do in the next chapter, one immediately spots the magical transformation of putative particle into insubstantial wave. Bohr will elaborate this observation into a new scientific philosophy called "complementarity" a few months later, which will cause even more uncertainty—among scholars, that is—than his younger colleague's conception.

With the publication of the uncertainty paper, the quantum revolution was essentially complete. (Bohr's contribution will be more in the nature of self-affirming ideology than radical new theory.) The classical view of reality, if not turned upside down, had at least suffered a most surprising transformation. According to Heisenberg, classical concepts such as position, momentum, time, and energy, had retained their meaning—indeed, precisely their original meaning—in the new scheme. But certain *pairs* of quantities had lost their role, although not entirely. They retain it up to "uncertainty" or "'imprecision" or "statistical error," which can be represented pictorially as follows (see Figure 4).

Think classically of the pair (Q, P), with Q denoting the position and P the momentum of a particle, as determining the Cartesian coordinates of a point in the plane. (Forget for the moment that these letters should each stand for three quantities, since physical space has three dimensions.) A

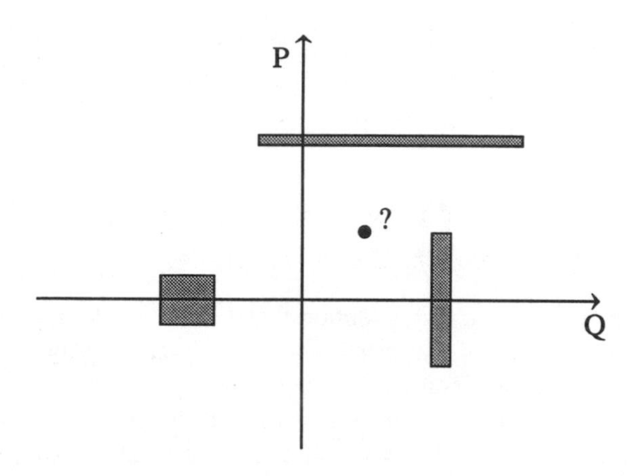

Figure 4

point in our (Q, P)-plane represents the complete classical description of a particle: it is the particle's classical *state*. For the sake of simplicity, set Planck's constant equal to one. Then here is the sense that the pair (Q, P) retains meaning, *pace* Heisenberg. (See Figure 1.) Draw any rectangle with sides parallel to the coordinate axes, of total area unity. The statement "the state of the particle is known to be (can be known to be? is likely to be?) in the rectangle" is meaningful. The rectangle therefore has meaning, and since it may become infinitely long and thin in either direction while retaining unit area, each coordinate individually has meaning. *But the point in the plane has no meaning.*

Now there is a possibility undreamed of in anyone's philosophy before Heisenberg!

6. Complementarity

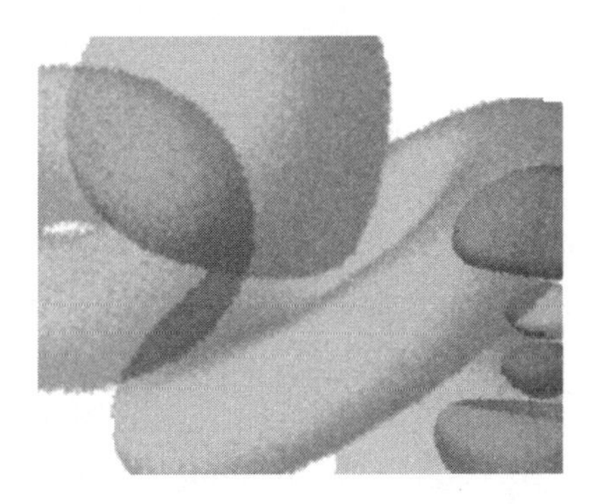

Dual pictures, dual language: *linguistic analysis* is the key to understanding quantum mechanics, Bohr told his protegé Heisenberg, shattering his hard-won vision of the microworld. The very words physicists use to describe reality constrain their knowledge of it, and scientists in every field will one day encounter this barrier to human understanding.

These were extraordinary assertions, unprecedented in the history of science. Before confronting them, it is best to review some cases of duality in particle physics. (Bohr's philosophical views from outside physics are discussed in Chapter 17, "Philosophies.") I begin with Bohr's objection to Heisenberg's interpretation of his gamma-ray microscope *Gedanken* experiment.

The complete analysis of this experiment, Bohr pointed out, exploits two inconsistent pictures of the nature of light. When gamma rays impinge on electrons, one treats the rays as a stream of little energy packets called photons. After a photon collides with an electron, the photon must pass through the optical train of the microscope if it is to contribute to forming an image of the electron. (At this point there is the complication that to form an image one needs an immense number of photons, either from many collisions with the same electron or from many repetitions of the experiment with different electrons. A troublesome point, but I will cheerfully ignore it, just as Bohr and Heisenberg did.) During this passage, however, the wave theory applies. Due to diffraction (a pure wave phenomena), the beam acquires a certain dispersion, or spread. The image is "fuzzed out" to a degree, resulting in uncertainty in the position of the electron perpendicular to the microscope's axis. The uncertainty in the electron's momentum, on the other hand, is derived entirely from particle mechanics. After colliding with the electron, the photon must be headed up into the optics of the microscope (see Figure 5), but that is all we can say for sure. Hence there is uncertainty about its momentum perpendicular to the axis, and consequently uncertainty about the direction of the "kick" imparted to the electron in that plane. Using Abbe's classical formula for diffraction and de Broglie's formula relating momentum and wavelength, one proves Heisenberg's law of uncertainty for this experiment.

After Bohr pointed out this striking appeal to two mutually contradictory pictures of light, Heisenberg capitulated. The reader may have a somewhat different reaction. How, you may ask, did the little billiard ball called "the photon" know to turn into a wave just in time to pass through the microscope? This is a good question—perhaps it is *the* question of 20th-century physics. Bohr waved a philosophical wand and caused this conundrum to vanish from physicists' thoughts, but before describing

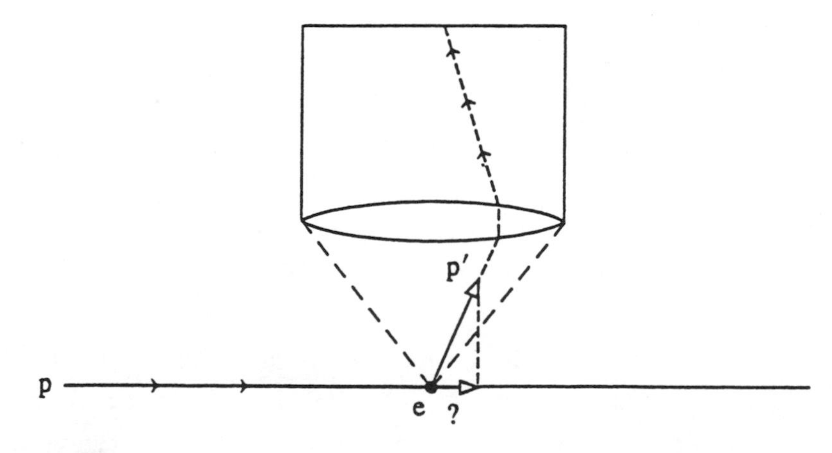

Figure 5

how he did so I introduce the exemplar of wave-particle duality: the "double-slit" experiment.

Richard Feynman thought this experiment to contain the essence of the quantum mystery. Consider a monochromatic beam of light incident on a screen containing two closely spaced parallel slits, as in Figure 6(a). On a second screen placed some distance behind the first, one observes a classical *interference pattern*: a pattern of light and dark formed by the alternation of "constructive interference" (an oxymoron meaning the summing of intensity for waves that arrive crest to crest and trough to trough) and "destructive interference" (the cancellation of waves arriving with crests partially or completely filling troughs). It was this simple demonstration that originally convinced 19th-century physicists that light was a wave phenomenon.

Suppose we insist on a "corpuscular" theory of light (as Newton did in opposition to Huygens); can we still understand the interference pattern? The problem arises when we close one of the slits. The pattern on the screen changes to that in Figure 6(b), and we see the difficulty: the two-slit pattern *is not the sum of two one-slit patterns.* Adding two one-slit patterns fails to generate all the "nodes," or bands of total darkness, in the two-slit pattern. Might the photons "know" to avoid the nodes when both slits are open? But each photon presumably passed through just one of the slits. If a photon passed through the slit on the right, it had no way

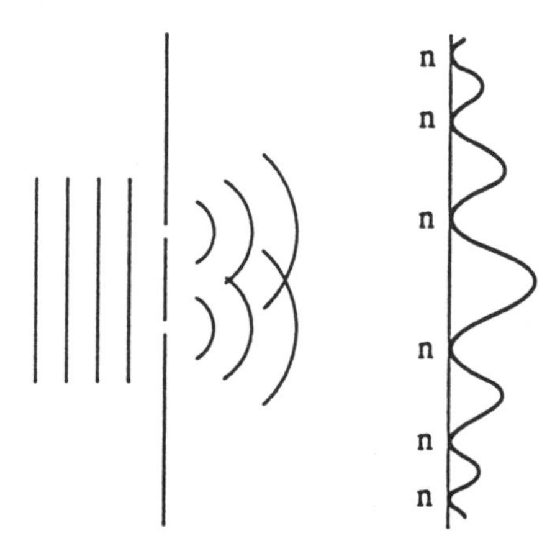

Figure 6 (a)

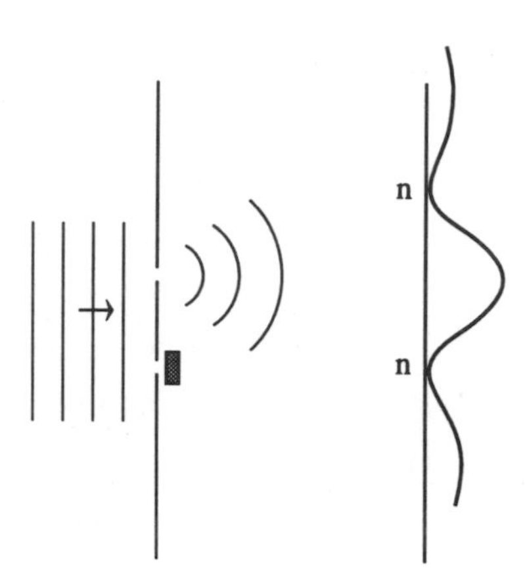

Figure 6 (b)

of knowing whether the slit on the left was open or closed. We have a paradox.

Perhaps the photons collectively imitate a wave. (This is not absurd, since sound waves show interference patterns; hence gas molecules have this capability.) But we have not yet faced the real quandary. Quantum mechanics predicts that the two-slit pattern appears *even if the intensity of the illumination is so low that the photons pass through the slits one at a time*. Experiments have verified this prediction.

Before coming to Bohr's interpretation of quantum mechanics, a few words on his writing style. The contrast with Einstein's prose is striking. Bohr wrote in a convoluted and awkward manner that rendered some of his passages almost unintelligible. (Bohr wrote in English, which was also the official language of his Institute in Copenhagen, so translation is not the problem.) Besides the stylistic problems, mysterious references to other disciplines (such as psychology) suddenly intrude into the argument, without obvious relevance or citations. Not surprisingly, Bohr's colleagues had considerable trouble understanding his views. Einstein complained that he could not achieve, after much effort, a "sharp formulation" of Bohr's complementarity principle. A student of complementarity, C. F. von Weizsäcker, worried for 25 years that he had misunderstood the basic idea. Finally, he presented his interpretation to the Master—and Bohr told him that he still had it wrong.

I now summarize, as best as I can, Bohr's "complementarity" arguments, based primarily on his expositions from the years 1948–1949. (Bohr first presented his views, which have come to be called the "Copenhagen Interpretation," in a speech at Como, Italy, in September 1927.) The first and most accessible idea is that "complementary phenomena," in which "contrasting pictures" are required, appear only in mutually exclusive situations. Experimental arrangements revealing one or the other are *physically* incompatible. Bohr may have thought of this dichotomy as discrete, "either–or," but later it was realized that the pictures shade from black to white through gray. Here is the simplest situation revealing this fact.

The double-slit experiment brings out the wave aspect of a photon or electron most clearly, while the Wilson cloud chamber (or its later refinements), in which electrons leave beautiful tracks like contrails of jet planes, best reveals particle behavior. So let us put a double-slit setup *inside* a cloud chamber and ask what happens. In a vessel containing air humidified to the saturation point, we place a source of electrons, an "electron opaque" screen with two slits, and a second screen, which fluoresces when struck by an electron. Question: Can we arrange matters so that we can *simultaneously* observe:

(a) the trajectories of the electrons, including which slit each passes through, and

(b) wave properties such as interference?

The answer is essentially no. We definitely cannot arrange to see sharp interference fringes (the alternating light and black bands, here representing zones where electrons do or do not strike the second screen) and simultaneously determine which slit the electron passed through. This follows from the finite size of the droplets and Heisenberg's uncertainty principle, as follows.

Assume for simplicity that in order to nucleate the formation of a droplet of size D, the electron had to pass somewhere through the space it occupies. We then know something definite about the electron's position in a direction perpendicular to the trajectory marked out by the droplets; see Figure 7(a). By the uncertainty principle, the electron "acquires" an

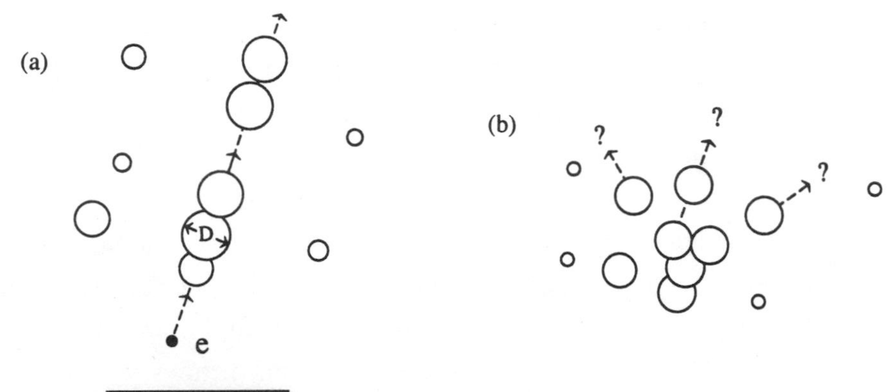

Figure 7 (a), (b)

"uncertainty," which we take here to mean a statistical spread, in the momentum in this direction given by

$$\Delta P_{\perp} \approx \frac{h}{D},$$

where $\Delta P_{\perp}$ is the uncertainty in momentum and h is Planck's constant. Now the condition that there be something remotely resembling a classical path is that $\Delta P_{\perp}$ should be much smaller than P, the electron's total momentum. Otherwise, the droplets will be scattered around so violently by chance variations in the momentum that we will be unable to "connect the droplets," as in Figure 7(b). From this we conclude that P should be much larger than h divided by D. Put another way, D should be much larger than h divided by P. Now h divided by P gives the wavelength of the electron, according to de Broglie. So we can state the requirement that there be any possibility of detecting classical-looking trajectories as follows: the droplet size must be significantly larger than the electron's de Broglie wavelength.

But there's the rub. To detect appreciable interference, the slit spacing should be about equal to the wavelength: if the slits are too far apart, the fringes in the interference pattern become so crowded together they cannot be seen. This forces the droplets to be larger than the slit spacing, dashing any hope of detecting which slit the particle passed through; see Figure 7(c). However, it is likely that one can see some interference

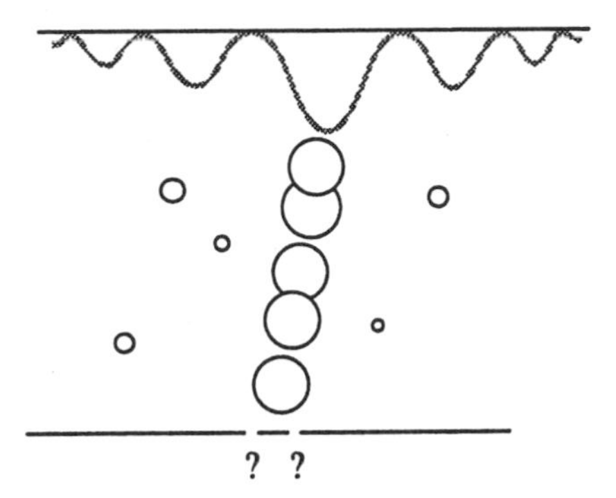

Figure 7 (c)

fringes and some aspects of particle behavior in this experiment, so the incompatibility is not absolute.

Note here how uncertainty is used to prove the impossibility of simultaneously observing wave and particle properties, while in Heisenberg's microscope experiment wave-particle duality served to bolster the uncertainty principle. This logical reciprocity is typical of arguments used to justify quantum mechanics. It is maddening: the principles chase each other around a circle, and one cannot get an independent understanding of either one.

So: incompatible experiments sometimes give rise to "complementary" pictures of phenomena. This part is indisputable. Now we come to the next most frequently repeated theme in Bohr's writing: "The account of all evidence must be expressed in classical terms." Bohr argues that all observations can be reduced to measurements made on macroscopic objects and that these are adequately described by classical mechanics. To this I do not object, but I would go further and point out that all measurements can ultimately be reduced to whether or not a pointer falls between two marks on a meterstick. One needs no concepts, classical or otherwise, to simply relate what one sees with one's own eyes in the laboratory. Concepts need enter only when we build theories accounting for the evidence of our senses, and then there are frequently several alternatives leading to the same predictions.

Perhaps Bohr meant that we can only *think* in classical pictures. I presume not, since the notion is absurd on the face of it. The modern-day student of mathematics shows no qualms when confronted with four dimensions, five dimensions, or infinitely many dimensions. Fractals are commonplace now, and Brownian motion for which "velocity" is meaningless troubles no one. The pictures that evolved between the 17th and 19th centuries certainly do not exhaust the powers of human imagination.

The force of Bohr's remark that all experiments must be described in classical terms, which he evidently thought of as a key point in the discussion, unfortunately eludes me. It also eluded Schrödinger, who pointed out in a letter to Bohr that when relativity theory replaced the older aether theory, "one did not say that the experimental results must be expressed afterwards as before in terms of the elasticity and density of the aether." Bohr responded that his main point was simply that measuring apparatus cannot be given a quantum mechanical description, a remark both surprising and trivial. It is surprising that a "complete" theory of atoms cannot describe a voltmeter, which after all is made of atoms, and trivial in that complex wave vectors do not appear on the dials of apparatus.

The next assertions are, unfortunately, even less transparent. Bohr proposes, as an interpretation of the uncertainty relations, another "complementarity," this time between "space-time description" and "causality." If one attempts to precisely locate a particle in space and time, this will demand an exchange of momentum and energy with the particle, as we saw in Heisenberg's microscope experiment. This exchange is "uncontrollable in principle," writes Bohr. The justification for this last statement lies in the "indivisibility" of quantum phenomena, due to the finiteness of the quantum of action (represented by Planck's constant). On the other hand, if we want to observe the "causal relations" embodied in the conservation laws, we will lose complete track of the location of the particles in space and time. By "causal relations" Bohr presumably refers to the fact that the conservation laws sometimes permit deductions, such as of energy or momentum exchanges in collisions, that would be impossible in a completely acausal universe. Bohr concludes that "space-time description and causality are complementary."

Because of the indivisibility of the quantum, Bohr says, one must accept the "impossibility of any sharp distinction between the behavior of atomic objects and the interactions with the measuring instruments which serve to define the conditions under which the phenomena appear." Consequently, "an independent reality in the ordinary sense can be ascribed neither to the phenomena nor to the agencies of observations." The word "phenomena" should even be restricted to what one directly observes in the laboratory—shades of Mach's "elements"!

The upshot of these unhappy circumstances is that any description of atomic phenomena more detailed than that given by quantum mechanics "is, in principle, excluded." But it is not a defect that quantum mechanics "amounts only to predictions, of determinate or statistical character, pertaining to individual phenomena appearing under conditions defined by classical physical concepts." In other words, quantum mechanics just provides a probability catalog for experiments described in everyday terms—and that is all one can hope for.

Complementarity, for Bohr, was not limited to atomic physics, but represented an epistemological lesson of general applicability. In psychology, it might express the disjunction between "thoughts" and "sentiments." In sociology, it might address the fact that prejudices inherent in one national culture cannot be grasped by the peoples of other nations. And, most interestingly, in biology it might explain the resort to "vitalistic" as opposed to physical explanations for behavior. (This had curious consequences; see Chapter 17, "Philosophies.")

Bohr provided the justification physicists needed to stop worrying and accept quantum mechanics as a completed theory—except perhaps for a future union with relativity and gravity. The psychological utility of this belief for the generation that witnessed the birth of quantum mechanics was undeniable. The end of history was at hand in atomic physics.

7.

The Debate Begins

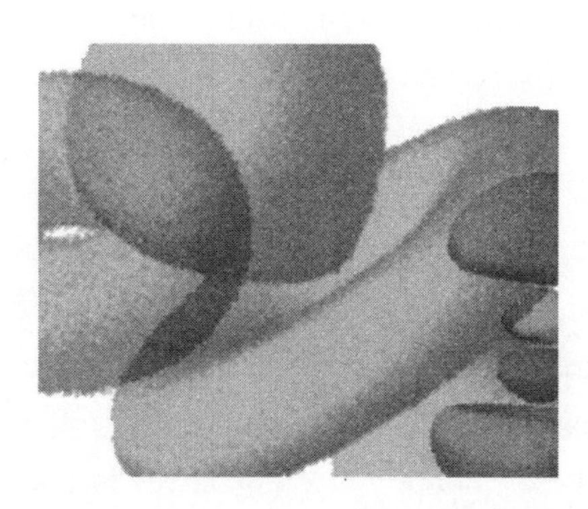

Quantum mechanics is very impressive.
But an inner voice tells me that it is not yet the real thing.
The theory produces a great deal but hardly brings
us closer to the secret of the Old One.

—Albert Einstein, in a letter to Max Born, December 1926

The Bohr-Einstein debate, the most interesting intellectual contest in science of this century, began in the late afternoon of October 24, 1927, at the Fifth Solvay Conference in Brussels, Belgium. (Einstein had not attended the meeting in Como a month before, at which Bohr first introduced complementarity.) The theorists Bohr, Born, de Broglie, Brillouin, Dirac, Einstein, Ehrenfest, Heisenberg, Pauli, Planck, and Schrödinger were in attendance, as well as the experimentalists Bragg, Compton, Madame Curie, and Debye. The eminent Dutch theorist H. A. Lorentz, the grandfather of relativity (after whom the theory's space-time transformations were named), chaired the sessions. The organizers had chosen "Electrons and Photons" as the official title of the conference, but a better one might have been: "Quantum Mechanics: What Does It

Mean?" As Einstein's opinion had become widely known, fireworks were expected.

Bragg and Compton opened the meeting by reviewing some experimental results. Then de Broglie got up to speak. He proposed an interpretation of Schrödinger's theory he called the "pilot wave." This involved "complementarity," but of a simpler kind than Bohr would elaborate that afternoon: a mere doubling of the hypotheses. *Both* waves and particles exist, de Broglie argued, with the waves leading or "piloting" the particles through space. If so, the relation of particle to wave is no more mysterious than that of a cork tossed into a stream to the water therein—except that corks tend to congregate at the troughs rather than the crests of waves. De Broglie had some formulas backing up his proposal, but it generated little enthusiasm among his peers, most of whom had come prepared to entertain more extravagant hypotheses. Dispirited by this reception, de Broglie abandoned his idea for Bohr's positivistic complementarity a year later.

Next Born and Heisenberg had their say, finishing with their declaration of victory. Schrödinger then lectured on problems in many-body theory and, remarking that for two or more particles his wave "flowed" through a multidimensional space, added his voice to the chorus against de Broglie. After some discussion in which Einstein did not participate, Bohr repeated the gist of his Como lecture of the previous month. There was more discussion, but again Einstein remained silent.

Only later, as the official discussions drew to a close, did Einstein finally rise to speak. "I have to apologize for not having gone deeply into the quantum mechanics," he began, implausibly. (Einstein's absorption in the quantum conundrum was well known. He once pointed out some residents of an asylum, remarking that those were the madmen not thinking about quanta, and on another occasion stated that he had thought a hundred times as much about the quantum problem as about general relativity.) Then he made two points. He contrasted the claim that Schrödinger's wave function gives a complete description of individual processes—the Copenhagen position—with the view that it represents an ensemble of systems (that is, a collection of possibilities). And he criticized the apparent recourse to a form of action at a distance. Einstein concluded:

> It seems to me that this difficulty cannot be overcome unless the description of the process in terms of the Schrödinger wave is supplemented by some detailed specification of the location of the particle.... I think that M. de Broglie is right in searching in this direction. If one works only with Schrödinger waves, the interpretation... contradicts the principle of relativity.

These issues would indeed form the core of the debate over the next three decades.

The discussion continued in private during meals at the hotel where the conference participants were put up. The interlocutors were Einstein, Bohr, Ehrenfest, and others of their generation; the youngsters—Heisenberg, Jordan, Pauli, Dirac—listened but were unworried, confident in what they had created. A pattern soon developed in which Einstein would make criticisms or propose thought experiments at supper, and Bohr would respond the next day at breakfast.

At first Einstein tried to refute the new doctrine on logical or physical grounds. He proposed a version of the two-slit experiment in which (it seemed) one could determine which slit the electron passed through and still observe a diffraction pattern. His arrangement is drawn in Figure 8 in the "pseudo-realistic" style later created by Bohr; the novel elements are the diaphragm containing a single slit placed in front of the one with the dual slits, and the tracks allowing the latter some freedom to move. Recall that the interference pattern appears even if the electrons pass through the apparatus one at a time, the pattern showing up only after a multitude of electrons have passed through by, for instance, making a time-lapse photograph of the screen.

Consider, said Einstein, an electron impacting near the center of the screen. Which slit did it pass through? If it passed through the left one, its momentum was changed from pointing to the left of center to pointing to the right, as in the figure. An interaction with the middle screen must have caused this perturbation. By the law of conservation of momentum, the screen must have suffered a corresponding recoil and should now be moving ever so slightly to the left. Alternatively, the electron might have

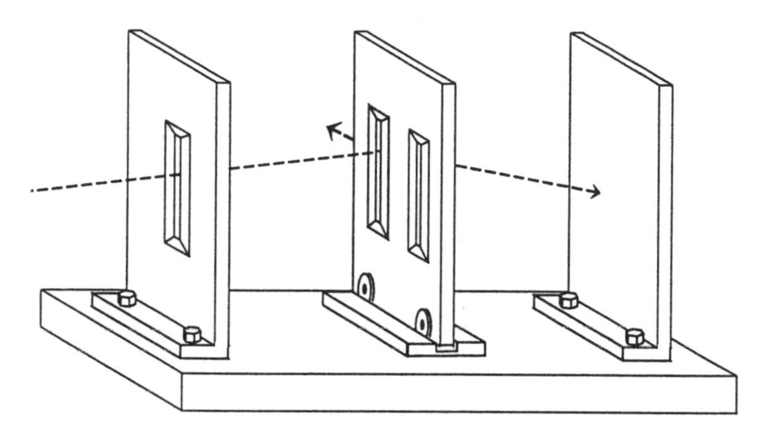

Figure 8

passed through the right slit, and the screen should be moving to the right. Einstein proposed measuring the middle screen's velocity with very great accuracy after the passage of the electron. Since the electron would be long gone at the time of this final measurement, the interference pattern could not be disturbed.

If Einstein's treatment of this experiment had been correct, the quantum revolution would have been derailed only a month after it gave the world a new metaphysics. Not only did Einstein have the complementarity principle in his sights, he was aiming to falsify Heisenberg's position-momentum uncertainty relation as well. If his experiment worked, one would know the perturbation in the electron's momentum with arbitrary accuracy. The slit itself could be chosen arbitrarily small, yielding precise information about the electron's position as it passed through. One measurement would fell both pillars of Copenhagenism.

Bohr was up to the challenge. One must perform *two* momentum measurements on the middle screen, he pointed out: one before the passage of the electron and one after. Otherwise, one could not compute the change. By Heisenberg's principle again, the first measurement must have produced an uncertainty in the screen's position. Therefore, the slits are at uncertain locations, and this additional uncertainty is just sufficient to cause the light and dark bands of the interference pattern to be superimposed, obliterating it—and saving the complementarity principle. The uncertainty in the slit positions also degrades knowledge of the electron's position as it passed through, restoring validity to Heisenberg's law.

Einstein, who was playing on their court, had to admit that Bohr was right. The new language of uncertainty and complementarity was proving surprisingly robust.

The antagonists met again at the next Solvay Conference, in 1930. Here, as Bohr put it later, "the discussions took quite a dramatic turn." Einstein proposed another thought experiment, aimed this time at the *time-energy* uncertainty relation, the twin of the better-known position-momentum one. This principle states that the product of the uncertainty of the time of an event, such as the emission or absorption of a particle by a material body, and the uncertainty in the change in energy of the body before and afterwards cannot be less than Planck's constant h. Einstein proposed a clock mechanism in a sealed box connected to a shutter arrangement. (See Figure 9, which reproduces a drawing prepared by Bohr in 1949 for his reprise of the debate with Einstein.) Light at low intensity fills the box. At an undetermined moment, the clock triggers the shutter to open for an instant, allowing at most one photon to escape. Now comes Einstein's clever idea: by carefully weighing the box before and after and exploiting that most famous of all equations:

$$E = Mc^2,$$

(E = energy, M = mass, c = the speed of light), which Einstein discovered a quarter-century before, one could compute the energy change of the combined system of box, mechanism, and light. Then by opening the box and reading the time elapsed on the clock one would also know, as accurately as the best Swiss technology will allow, the time at which the photon was emitted.

Bohr was taken aback. As his junior colleague Rosenfeld recalled 30 years later,

> ... he did not see the solution at once. During the entire evening he was extremely unhappy, going from one to the other telling them that it couldn't be true, that it would be the end of physics if Einstein were right; but he couldn't produce any refutation. I shall never forget the vision of the two antagonists leaving the club: Einstein, a tall,

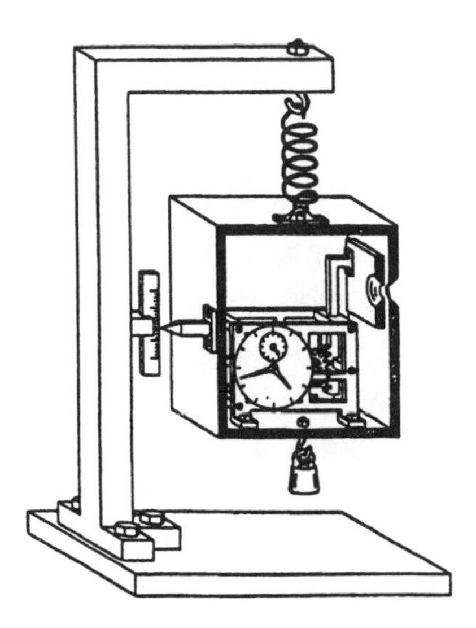

Figure 9 Reprinted from *Albert Einstein: Philosopher-Scientist*, Volume VII, P. A. Schlipp, ed., with permission of Open Court Publishing, Chicago, Illinois, 1949, 1951, 1969, and 1970.

> majestic figure, walking quietly, with a somewhat ironical smile, and Bohr trotting near him, very excited.... The next morning came Bohr's triumph.

Bohr's solution to the puzzle was as clever as Einstein's construction of it. Since Einstein based his argument on an equation from the theory of relativity, Bohr naturally based his refutation on that theory as well—but, remarkably, Bohr used a result of the *general* theory of 1915 (Einstein's theory of gravity), not the *special* theory of 1905 (from which $E = Mc^2$ was derived). The effect that saves the time-energy uncertainty principle, according to Bohr, is the slowing of a clock's rate in a gravitational field.

One of the predictions of the special theory of relativity was the slowing of all clocks moving at the same speed by a uniform factor, so that one could say that *time itself* slowed down in a moving reference frame. The general theory made an even more surprising prediction: the same uniform slowing occurs in a region of space near a massive body, relative to a region less subjected to gravitation. Thus a clock on Jupiter ticks more slowly than an identically constituted clock on Earth. A clock at sea level ticks more slowly than one on a mountain top. And so forth. This time-slowing effect reflected the explanation of gravity in Einstein's theory as a warping of *spacetime*, not just of space, a common misunderstanding. Since a light wave, in Maxwell's conception, represents a periodic oscillation of the electromagnetic field, light generated near the surface of the sun should have a longer period of oscillation, and therefore a longer wavelength, than a wave generated by the same causes on the Earth. Longer wavelength means redder light. Einstein based one of three proposed tests of his new theory on this gravitational "red-shift." (It proved too small to be detected in the sun, but was soon observed in massive stars such as Sirius. Thanks to continuous improvement in technique, it has since been seen in terrestrial physics departments.)

Omitting the mathematical details, here is the gist of Bohr's argument. Initially, the pointer on the box will be very accurately zeroed on the scale by use of carefully selected weights. Later, long after the photon has departed but before opening the box to read the clock, one reweighs it by adding weights, attempting to return the pointer to the equilibrium position. There will necessarily be a certain latitude in the measurement of the position of the pointer, determined by the finest division on the fixed scale. By Heisenberg's first uncertainty principle, there will be an associated latitude in the momentum of the box in the vertical direction. This will cause an additional vertical motion, which will have to be overcome by adding or subtracting weights during the weighing procedure, generating a latitude in the measured weight of the box. During this procedure, the box is certain to move by a slight but definite amount vertically. By Einstein's general theory of relativity, the clock inside will slow down or speed up by an amount involving the distance moved, the strength of Earth's gravity, and the square of the speed of light. Computing the effect on the clock rates, Bohr showed that the time-energy uncertainty principle is restored.

Another victory for Bohr, this time sweetened by irony: Einstein's greatest creation has been used to deflect his latest attack on the Bohr-Heisenberg philosophy. But there are multiple ironies in this episode. Einstein had taken exception to Bohr and Heisenberg's frequent comparisons between the impossibility of sending signals faster than light as the basis of the special theory of relativity, and the impossibility of simultaneously measuring the position and velocity of a particle as the basis of quantum mechanics. (When Philipp Franck suggested to Einstein in 1930 that the Bohr-Heisenberg philosophy "had been invented by you in 1905," Einstein replied: "A good joke should not be repeated too often.") Einstein hoped to replace this putative parallel with a direct clash, in which relativity would refute quantum mechanics. The outcome was the reverse of that intended, but a final irony was in store for these two interlocutors: The kernel of a much stronger argument resides in this clock-in-a-box *Gedanken* experiment.

In passing, note Bohr's choice of the seemingly innocuous term "latitude" for the more metaphysical expression "indeterminacy" (which he uses at some points of the manuscript) or the psychological formulation "uncertainty" (which he avoids). Here we have yet a fifth descriptive term for this elusive concept. The use of the neutral, colorless word "latitude" throws a tranquilizing blanket over the ambiguities lurking in the uncertainty principles; if Bohr had used "uncertainty," the reader might have wondered how one's lack of knowledge can set a box in motion. This transformation of uncertainty into activity is characteristic of arguments based on Heisenberg's laws.

One more remark on this stimulating exchange: some thinkers later charged Bohr with cheating, bringing in general relativity even though Einstein only exploited the special theory. But this is unfair to Bohr. Einstein used weight in his demonstration, which is meaningless without a theory of gravity. It was only natural that Bohr appealed to the best theory of gravity ever devised to analyze the experiment. He could hardly be blamed if it had the effect of hoisting Einstein by his own prediction. (I suspect that Einstein overlooked the appeal to the position-momentum uncertainty relation again, just as in 1927, and so missed the relevance of his clock-slowing effect.)

After 1930, Einstein abandoned the attempt to find a logical contradiction in the new theory. He proposed Schrödinger and Heisenberg for the Nobel Prize in 1931, writing to the Nobel Committee that quantum mechanics "undoubtedly contains a piece of the ultimate truth." But he was not ready to concede defeat in the contest with Bohr.

8.

The Impossibility Theorem

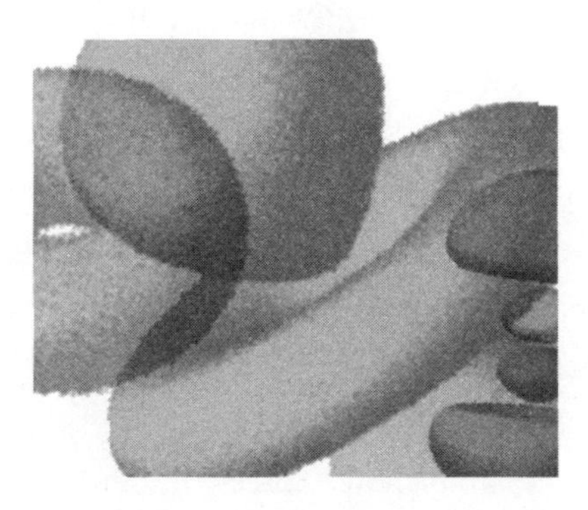

"Nothing takes place without a sufficient reason," wrote the German philosopher and mathematician Gottfried Leibniz in the 18th century; "that is to say, nothing occurs for which one having sufficient knowledge might not give a reason... why it is as it is and not otherwise." When 19th-century physicists invoked chance in their statistical theories of matter, it cast no cloud over the principle of sufficient reason; probability is useful, but causal determinism rules the universe. So it was most surprising when, in 1932, a young Hungarian mathematician announced a *mathematical proof* that determinism is dead.

John Neumann was born in 1903 into a middle-class Jewish family in Budapest. His father prospered as a banker, to the extent that in 1913 he was offered the opportunity of ennoblement by Franz Joseph, emperor of Austria-Hungary. (The government hoped by offering titles to selected industrialists and bankers to cement the alliance of the nobility with the bourgeoisie.) Thus "von" was added before the family name. A year later John—the oldest of three sons—entered the Lutheran Gymnasium in Budapest. A few months into the term, the mathematics teacher informed the von Neumanns that their eldest was a mathematical prodigy.

The gymnasium teacher suggested that a tutor from the university be obtained for the boy. This was done, and soon the small world of Hungarian mathematics was aware of a budding genius in their midst. Eugene Wigner, another of the brilliant scientists Budapest produced at this time, recalled his friendship with von Neumann. "We often took walks and he told me about mathematics and about set theory and this and that.... He was inexhaustible on such occasions.... He was phenomenal." Wigner, who would later win a Nobel Prize for his treatment of symmetries in quantum mechanics, remarked that he realized for the first time the difference between a really first-rate mathematician and someone like himself.

After completing high school in Budapest, von Neumann considered following his father into business, but his developing talent, encouragement by established mathematicians, and the political turmoil in Hungary—the von Neumann family had fled to Vienna after the Bela Kun socialist faction took over in Budapest—convinced him to turn to academic life. When a second upheaval brought a right-wing government back to power, John enrolled in the University of Budapest but soon left for Berlin and Göttingen, returning only for examinations. In Berlin he heard Einstein lecture on statistical mechanics, and in Göttingen Hilbert taught him the axiomatic approach to mathematics.

David Hilbert (1862–1943) was the Grand Old Man of German mathematics. A fervent advocate of the axiomatic approach bequeathed to the world by Euclid, Hilbert had tried to find foundations for mathematics that would secure it from uncertainty or paradox. The young von Neumann naturally joined this effort (on which Bertrand Russell had labored long) and made laudatory contributions, but the program was soon derailed by Kurt Gödel's incompleteness theorem. In 1900, Hilbert had stated 23 problems for mathematicians to tackle in the next century, and many a prize-winning paper would later begin, "On Hilbert's *n*th problem...." The sixth problem was to find axioms for physics that would make possible the certainty that rigorous argument provides in mathematics.

In 1925, von Neumann took a degree in chemical engineering from the ETH in Zürich—even geniuses need a fallback position—and then finished his doctorate in mathematics at the University of Budapest. It was 1926—the year of the quantum revolution.

Given the timing, it was probably inevitable that the brightest young mathematician in Europe would write a book about quantum mechanics. Von Neumann heard Heisenberg lecture on matrix mechanics in Hilbert's seminar in Göttingen and then began working with Hilbert on putting the field on a sound mathematical footing. (This was a collaboration between men more than 40 years apart in age!) Von Neumann would find the tools he needed in Hilbert's earlier construction of a kind of geometry possessing an *infinite number* of dimensions.

I remember reeling when a classmate uttered that last phrase in college, but it is not as shocking as it sounds. Although the attempt to visualize an infinity of perpendicular axes all meeting at a point should be avoided as conducive to mental strain, the notion arises naturally in a familiar setting. Hilbert abstracted it from Fourier series, sums of trigonometric functions that can express a chosen function. (The decomposition of the sound in an organ pipe into a fundamental tone, a first overtone, and so on, is one example of Fourier series.) In a natural sense these vibrations are "perpendicular" to one another, and the "space" of all such functions forms an infinite-dimensional generalization of Euclidean geometry. (For a little amplification, see the Notes.)

Working with Hilbert, von Neumann explored the symmetries of what is now called "abstract Hilbert space." Hilbert had studied how to set up and solve equations in his novel "space"; these equations typically involve infinitely many variables. The equations of Heisenberg, Born, and Jordan seemed to fall in this arena, but the Göttingen mathematicians quickly spotted a flaw. The matrices of these quantum theorists were too far from the finite case—in imprecise language, they were "too infinite"—to be covered by Hilbert's prior work. So von Neumann extended Hilbert's theory to include these "unbounded operators." Out of the need to understand quantum mechanics, a new branch of mathematics was born. (Not that physicists necessarily noticed. When the mathematician K. O. Friedrichs remarked to Heisenberg after the war that von Neumann's drawing of the distinction between the various kinds of operators meant mathematicians had finally paid him his due, Heisenberg said, "Eh? What's the difference?")

In 1932 Von Neumann's book *The Mathematical Foundations of Quantum Mechanics* appeared. It was the second text attempting to give quantum mechanics a proper mathematical form; the first was Paul Dirac's *The Principles of Quantum Mechanics*, published two years earlier. Von Neumann made a few testy remarks about his English competitor's book, especially concerning his use of a function taking the value infinity at one point, now known as "Dirac's delta function." (Von Neumann's attack was later vitiated when mathematicians built a useful theory of such improper functions.) Then he built the rigorous mathematical machinery needed to prove Schrödinger's and Heisenberg's theories equivalent, to define the "spectrum" or energy levels of an atom, and many other theorems. Given the level of rigor in von Neumann's book and the physicist's traditional indifference to axiomatics, few probably pushed through it cover to cover (Dirac's was an easier read), but by osmosis, as it were, many of von Neumann's results became widely known. Together these two books reassured physicists that there was no mathematical contradiction in their new fundamental theory.

The most important point on which physicists needed reassurance later became known as the Measurement Problem. How did the everyday world of demonstrable facts—of pointers on dials that point to something, and of billiard balls that are never seen on both sides of the table at once—emerge from the nebulous mix of uncertainty and duality that was quantum mechanics? Here von Neumann's treatment was seminal. Eventually the Measurement Problem was recognized as central in the debate, but to avoid a lengthy digression I defer further discussion until Chapter 16, "Paradoxes," and later chapters.

Chapter 4 of von Neumann's book contained the celebrated proof that determinism—and with it classical physics—was overthrown. Now it is reasonable to ask how a mathematician might prove such a thing. Weighing the evidence is the statistician's *métier*, not the mathematician's. Mystical insight is not normally permitted as a component in mathematical proofs. Hence the only route open to a mathematician is to analyze theories on the basis of their *assumptions*. Von Neumann wanted to prove an "impossibility theorem," sometimes called a "no-go" theorem: one whose conclusion is that certain kinds of theories cannot explain the phenomena of interest. So he listed all assumptions any reasonable theory of atoms should possess and then derived a conclusion that seemingly ruled out the competition.

Before discussing the pitfalls in this scheme, here is, in outline, von Neumann's proof that deterministic theories can never explain atoms. I will use a bit of symbolic notation, but it should cause no confusion and will speed up the exposition. Von Neumann first notes that all observable quantities in quantum mechanics—position, momentum, energy, and so on—are represented by certain kinds of matrices (Heisenberg's square arrays) which are called *Hermitean*, after the 19th-century French mathematician Hermite, who used them in a study of analytic geometry. These matrices, which I will denote simply by capital letters $H, K, \ldots,$ encode all possible values of the corresponding observable in a sense I will not further explain; one can just think of them as convenient labels. I denote the observable corresponding to a matrix H by $O(H)$. We need only one more fact about these Hermitean matrices: adding two together in the obvious way (by adding the numbers in the arrays) yields another.

Now a basic observation of atomic physics is that, when measuring an observable of a quantum system, one often gets different values on different "runs." This may happen even if the system is prepared the same way each time. Atoms behave randomly—although this is not yet von Neumann's theorem, since coins and dice do so also, without casting doubt on the principle of sufficient reason. Because of the randomness, it is useful to make many measurements of the atom in the same condition—or, as physicists say, in the same "state"—and compute a statistical average, just as the weather service does for noontime temperatures in the month of February. Traditionally, physicists use two angles back to back, *viz* $\langle \cdot \rangle$, to denote statistical average. Putting these notations together, the average of the energy observable is denoted

$$\langle O(H) \rangle$$

where H is the matrix corresponding to the energy. For instance, if energy were measured to be 2 units on 10 occasions and 3 units on 10 others, its average would be approximately $2 \times 10 + 3 \times 10/20 = 2.5$ units. By "$\langle O(H) \rangle$" is meant an average from a very long series of trials.

Von Neumann made the following innocuous-looking assumptions about *any* theory of atomic phenomena, which of course are valid for quantum mechanics:

1. There are at least as many observables as Hermitean matrices.
2. If one adds the matrices, one adds the averages. Symbolically,

$$\langle O(H + K) \rangle = \langle O(H) \rangle + \langle O(K) \rangle.$$

The first assumption just assures us that the theory has as many observables as quantum mechanics and so explains as much about atoms. The second seems to follow from the very definition of an average as a sum divided by the number of observations. From these innocent assumptions and some brilliant algebra, von Neumann drew a remarkable conclusion: *there are no dispersion-free states.*

By a "dispersion-free state" von Neumann meant a state of a system—an atom, say—that displays no randomness. Quantum systems have no dispersion-free states. For any state there is always some observable which fluctuates, no matter how much care is taken in preparing the state initially. For example, for a particle this follows from Heisenberg's uncertainty principle (in its statistical guise): if you fix the position, the momentum becomes completely random, and vice versa. (For specialists in the audience who just hooted: if you approximately fix the position, the momentum develops a large variance.) By contrast, a pair of dice exhibits a nondispersive state when sitting still on the table.

Von Neumann interpreted his conclusion as follows:

> It is therefore not, as is often assumed, a question of re-interpretation of quantum mechanics—the present system of quantum mechanics would have to be objectively false in order that another description *of the elementary process* than the statistical one be possible.

(My emphasis.) Von Neumann thought he had proved that, in modern terminology, the world is a *stochastic process*.

This last phrase refers to a series of events, such as idealized coin flips or dice throws, in which *genuine randomness* comes in with each successive event. In a stochastic process no amount of information suffices to predict subsequent events with certainty. To avoid confusion this kind of process must be carefully distinguished from two other types of "random processes"; see Figure 10. Consider a billiard ball moving on a perfect billiard table, but whose initial position and velocity are unknown and hence "random." (Here I am confounding many interpretations of the term "random," in particular the subjective and the frequency interpretations. I do not wish to address this debate here, which has been raging since the 17th century.) This sort of random-but-predictable process had appeared a century earlier, in Maxwell's kinetic theory. Now put a convex obstacle such as a weighted tire in the center of the table. A tiny initial perturbation in the ball's velocity will result in large deviations at later times, a situation called "deterministic chaos," or simply (and inaccurately) "chaos." Such unstable processes have been much studied in recent times by mathematicians and others. For both these deterministic systems, one obtains a dispersion-free state by precisely specifying the ball's initial conditions. Von Neumann thought his theorem ruled out these kinds of explanations for atoms.

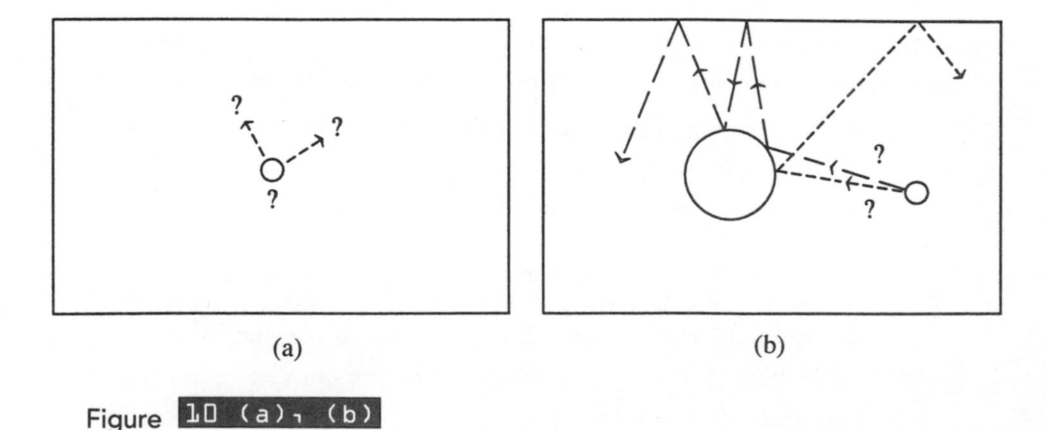

Figure 10 (a), (b)

Curiously, some people thought von Neumann had abolished the real world! Although such a philosophical feat has been attributed to Prince Gautama of India (563?–483? B.C., called the Buddha), it would have been a stunning performance for a mere mathematician. The notion arose from von Neumann's use of the phrase "hidden variables theories" for the explanations he thought he had ruled out. Since quantum mechanics did not give a realistic description of atoms, went this reasoning, and since von Neumann had ruled out anything hiding behind the quantum facade, nothing was left. Although Mach might have been pleased, von Neumann preferred to think he had abolished determinism—a comforting thought for those who believe themselves possessed of free will. (Perhaps someday an advanced version of that other invention of von Neumann, the electronic computer, will retort that people are also machines.)

Von Neumann's proof was quoted by physicists and philosophers for 30 years; occasionally it is even today. Part of the difficulty in discovering the error was that the theorem, *qua* theorem, was correct. (I doubt that von Neumann ever made a simple mathematical blunder.) The problem lay not in the proof, but in von Neumann's interpretation of his theorem. An ex-student of the mathematician Emily Noether who had switched from mathematics to philosophy first noted the circular form of von Neumann's reasoning. Grete Hermann's analysis might have exploded von Neumann's claim to have abolished determinism in 1935, but unfortunately she published in an obscure philosophy journal. When David Bohm constructed in 1952 exactly the kind of explanation of atoms—classical and deterministic—that von Neumann thought he had ruled out, the game should have been over. (See Chapter 10, "The Post-War Heresies.") But somehow it was not. The end came, or should have come, in 1965, when John Bell found a classical model of an electron's "spin" so simple it brought the error into plain view.

Von Neumann submitted to the discipline of the axiomatic method, and it is the surest route to truth the human race has yet devised. But in his "impossibility theorem" von Neumann indulged in reasoning going beyond the straightforward deduction of theorems *about a single theory* by the rules of logic. His theorem required an analysis, in a sense, of "all possible theories" (or at least all reasonable theories) of the phenomena at hand. There are difficulties with such an investigation.

The first hurdle is to get a complete understanding of the given theory. One naturally tries to draw up a list of all the assumptions made, but this can be surprisingly difficult; given such a list, checking the logic of the theory can be child's play by comparison. Some assumptions are there on the printed page—for example, that energy can be neither created nor destroyed in an exposition of thermodynamics. Others, such as the laws of probability or the rules of the calculus, are implicit but innocuous. But there may be yet other, unstated presumptions that are so deeply ingrained in one's world view or culture that no one thinks to question them. Newton's assumption of an Absolute Time had this character before Einstein challenged it in 1905; in later chapters we shall encounter many more instances of such "hidden assumptions."

A second pitfall is circular reasoning. The danger here is that the hypotheses of one's favorite theory may start to resemble unavoidable presumptions, or necessities born of the phenomena. Von Neumann's attempt to rule out determinism foundered on this rock, for his second axiom was neither. The essence of the quantum predicament, as Bohr and Heisenberg repeatedly emphasized, is that measurements may be physically incompatible. In the case of von Neumann's second axiom, incompatible apparatus may be required to measure the first two observables, with yet a third needed to measure the one corresponding to the sum of their matrices. When this is realized, the equality assumed in that axiom becomes (as John Bell pointed out in 1965) a surprising property of quantum mechanics, and by no means a necessity in any alternate theory. (Von Neumann mentions physical incompatibility in his book, so it is hard to understand his ignoring it here.) Von Neumann ruled out the competition by hypothesis, not by analysis.

Ignoring the influence of the apparatus is worth further discussion, since it is so easy to fall into this trap. All of the "impossibility theorems" proved in this century founder on just this point; see the Notes. In 1965, David Bohm and Jeoffrey Bub constructed a realistic model of a "quantum apparatus"—a measuring device yielding the output predicted by quantum mechanics—which had "hidden internal states." (See Chapter 18, "Principles.") Wigner recalled asking von Neumann about such hidden variables in the apparatus at the time, but von Neumann ruled them out, stating that enough measurements would "fix" these variables. But 10^{24} or so atoms are needed to make a macroscopic apparatus, while one atom makes a quantum system. Might not "fixing" the apparatus state take (say) 10^{20} additional measurements, each lasting one second? (Note: 10^{20} seconds is about 10^{12} years.)

Von Neumann's "impossibility theorem" cannot properly be called a mistake; indeed, it helped—after much additional work by others—to illuminate how quantum mechanics differs from rival theories. Yet his conclusion was undeniably in error, and one has to ask how a man who helped create modern mathematical logic, operator theory, and game theory, and who is sometimes called "the father of the electronic computer," could have fostered such a misunderstanding. Perhaps proving one's world view as certain as that $1 + 1 = 2$ is a task beyond even the reach of genius.

9.

EPR

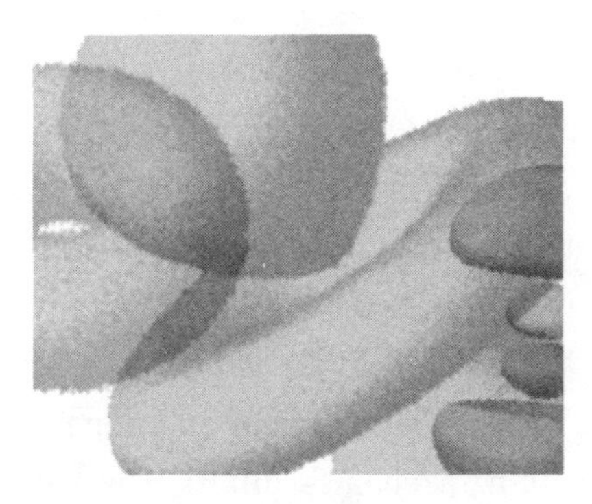

The third round of the Bohr-Einstein debate was conducted by way of the journals rather than in person. Five years after Bohr's victory at the Sixth Solvay Conference, Einstein, working with two younger colleagues, Boris Podolsky and Nathan Rosen, finally achieved a sharp formulation of the argument that quantum mechanics is an *incomplete description* of reality. They published in the American journal *The Physical Review*, in 1935. As one expects in a paper bearing Einstein's name, the argument is simple but very clever. An improvement by Schrödinger speeds up the exposition.

Einstein, Podolsky, and Rosen—now universally abbreviated "EPR"—were going to criticize the account quantum mechanics gave of reality. So, striving for precision, they actually defined the term. Here is their careful definition of an "element of reality":

> If, without in any way disturbing the system, we can predict with certainty (i.e., with probability equal to unity), the value of a physical quantity, then there exists an element of physical reality corresponding to this physical quantity.

The emphasis on not disturbing the system under observation was of course designed to foil the expected objections from Bohr. EPR probably thought Bohr believed in properties being created by measurements, or measurements causing perturbations in the system measured. But Bohr had gone beyond all that.

For their *Gedanken* experiment, EPR considered two particles which are subjected to various measurements in two laboratories separated by a great distance. In either laboratory an experimenter carries out one of two incompatible measurements on a particle: either a very accurate measurement of position, or an equally accurate one of momentum. Now, according to the rules of quantum mechanics, the *difference of the positions* and the *sum of the momenta* of two particles are "compatible" quantities, without any "reciprocal uncertainty" as holds for a single particle's position and momentum by Heisenberg's law. (This was Schrödinger's improvement; EPR invoked collapse of the wave packet in a measurement. For the latter, see Chapter 16, "Paradoxes.") Thus we can arrange matters so that the positions of the particles differ by a known quantity, and, simultaneously, the sum of their momenta is zero (which just means that they are moving in opposite directions). The difference between positions will grow with time, but the important thing is that the uncertainty does not grow with it.

EPR next demonstrate that each particle possesses a definite position and momentum. Consider the particle in laboratory A. In laboratory B, one can measure *either* the position *or* the momentum of the particle there, but not both. Using the sum and difference laws quoted above, one deduces *either* one *or* the other for the A particle. Since such determinations are possible, and since no influence can propagate sufficiently rapidly between the two laboratories to disturb the first particle in any way, both quantities must have simultaneous reality. But quantum mechanics denies this, or at least refuses to state what they are. Therefore, quantum mechanics is not a complete theory of the phenomena.

This counterrevolutionary thrust "came down upon us as a bolt from the blue," recalled Bohr's assistant Léon Rosenfeld. Bohr rushed to find a refutation. After a few months, his reply was ready; it appeared in the same year and in the same journal. According to the established mythology,

Bohr's response definitively shot down the reactionaries' argument. For instance, Rosenfeld wrote 30 years later:

> Einstein's problem was reshaped and its solution formulated with such precision and clarity that the weakness in the critic's reasoning became evident, and their whole argumentation, for all its false brilliance, fell to pieces.

I first read the EPR paper and Bohr's response in graduate school, in the 1970s. EPR's essay seemed terse and well-reasoned, I recall, but Bohr's prolix and obscure. (I have since heard from senior colleagues that Bohr's locution, as opposed to his writing, was generally persuasive—although recalling the argument later was another matter!) Others who read this exchange at a late date have had the same reaction. John Bell, who proved his understanding of the EPR puzzle by solving it in a way that both Einstein and Bohr missed, once remarked that he could not understand any of Bohr's criticisms of the EPR paper.

Bohr begins his essay with a long reprise of uncertainty and complementarity. Then he describes EPR's *Gedanken* experiment and at a crucial juncture produces this passage, which I reproduce in its entirety:

> ... The criterion of physical reality proposed by [EPR] contains an ambiguity as regards the meaning of the expression "without in any way disturbing the system." Of course, there is in a case like that just considered no question of a mechanical disturbance of the system under investigation during the last critical stage of the measuring procedure. But even at this stage there is essentially the question of *an influence on the very conditions which define the possible types of predictions regarding the future behavior of the system.* Since these conditions constitute an inherent element of the description of any phenomena to which the term "physical reality" can be properly attached, we see that the argumentation of the mentioned authors does not justify their conclusions....

(The emphasis is Bohr's.) That same year Bohr wrote a letter to the British journal *Nature*, including this mysterious sentence as the sole counterargument to EPR—proving that he thought it alone did the job.

The meaning of Bohr's words in this passage eludes me. (Bell also expressed his bafflement over it.) Curiously, a decade later, Bohr admitted he could not understand it either! In his long review of his debate with

Einstein, published in the Einstein Festschrift volume of 1949, below a reproduction of the selection quoted above, Bohr wrote, " Rereading these passages, I am deeply aware of the inefficiency of expression which must have made it very difficult to appreciate the trend of the argumentation...." I cannot accept Bohr's excuse. Both Bohr and, later, Bell singled out this one paragraph for examination not because it is an example of sloppy exposition, but because it is the clincher in Bohr's argument. If it is impossible to make sense of it, Bohr has no case.

There were other grounds to object to EPR, although Bohr likely rejected them on ideological grounds. For instance, he might have argued that the knowledge from the second laboratory is useless. I make this argument first for a single particle.

Sometimes the uncertainty principle is described by the words: "You cannot know both the position and the momentum of a particle at a single instant in time." But this is incorrect; there are many cases where you can know both. Consider, for example, the "single-slit experiment," which Bohr used to illustrate uncertainty and wave-particle duality in his 1949 article. A wave representing a particle passes through a screen with a single slit. Approaching the screen from one side, the wave fronts are straight, but on the other side they fan out into a circular pattern because of refraction at the slit, see Figure 11. This spreading is the source of the uncertainty in the particles' momentum parallel to the screen. Now suppose the particle hits and makes a tiny flash on a second screen. Assuming the particle moves classically in a straight line, from simple geometry one can work out what its momentum parallel to the screen must have been. (The horizontal and vertical components of momentum, thought of as arrows (vectors), must make a similar triangle to that made by the screen, the slit, and the point where the particle hit. Hence the ratio of the momentum components is the same as the ratio of the sides in the physical triangle. The horizontal momentum can be measured beforehand to arbitrary accuracy, without generating any new uncertainty in the vertical component.) Knowing this and that the particle passed through the slit, which can be made arbitrarily small, violates the naïve statement of the uncertainty principle.

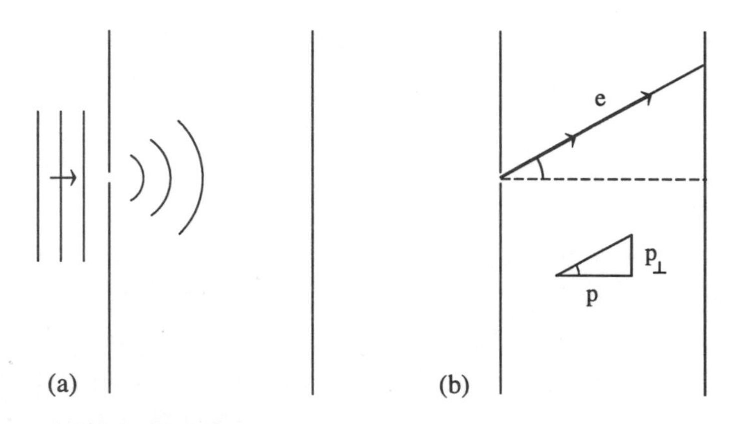

Figure 11 (a), (b)

However, this knowledge obtained by "retrodiction" is useless for later experiments, since the particle has been absorbed by the screen. "It is a matter of personal belief," wrote Heisenberg in 1929 about similar cases, "whether such a calculation concerning the past history [of the particle] can be ascribed any physical reality or not." Bohr might have made the same point in reply to EPR. Knowing that laboratory B will eventually provide information about the particle's position in laboratory A, the A experimenter is free to measure its momentum directly. But this will so jostle the A particle that the position knowledge communicated from B will become useless. Bohr probably did not make this "useless knowledge" objection because he did not want to admit that EPR were right about incompleteness. Also, as we saw in Chapter 6, "Complementarity," Bohr did not like the idea of measurements "perturbing" what was "really there." Like Mach arguing against Boltzmann, Bohr wanted to refute EPR on *epistemological*, not physical, grounds.

Strangely, by the time of his 1949 summing up, Bohr had not found a better argument against EPR. Considering the mythic status this exchange attained in the minds of physicists as the final triumph of Bohr over Einstein, the brevity of Bohr's remarks is astonishing. He states petulantly that "we are dealing here with problems of just the same kind as those raised by Einstein in previous discussions," which is not true; accuses EPR again of ignoring an "essential ambiguity" in their criterion of reality; and closes with an observation of underwhelming force: "As repeatedly stressed, the principal point here is that such measurements demand mutually exclusive experimental arrangements."

But EPR had invented their argument precisely to get around this objection. Despite the judgement of the physics community, Bohr found no refutation of either EPR's assumptions or argument; he begged the question.

That same year Schrödinger invented a nice metaphor for the behavior of a particle in quantum mechanics. (It comes from the paper with the more famous cat metaphor; see Chapter 16, "Paradoxes.") He compared it to that of a diligent but nervous schoolboy faced with an examination. The boy studies hard and so is always able to answer the first question asked of him correctly, no matter what question is chosen from a standard list. But then he gets nervous or pooped out and cannot answer a second question. Yet would not everyone admit that he must have known all the answers when he first walked into the room? Adding a second student in a second examination room who studied with the first, we could even find out, as in EPR, how the boy *would have answered* that second question.

The EPR paper was Einstein's last important contribution to the debate, except for a summary written a year later (and a short response to Bohr in 1949). Einstein fled Germany for America in 1933 and settled in Princeton. In 1936, he wrote a long memoir entitled "Physics and Reality," which was published in German but accompanied by an English translation, in the *Journal of the Franklin Institute* of Philadelphia. It is somewhat out of place in this practical Yankee journal. The long essay of the German abstract thinker pondering the nature of reality is followed by an article entitled "Vacuum Pumps and Pump Oils, Part II. A Comparison of Oils" and a report by the National Bureau of Standards on the penetration of moisture into masonry walls. (I wonder if the ex-patent clerk, third class, reflected on this whimsical juxtaposition?)

Einstein first reviews the development of classical ideas, of atomic physics, and of relativity theory. These are fascinating sections, but outside our purview. He concludes with a section entitled "Quantum theory and the fundamentals of physics." After preliminary laudatory remarks:

"Probably never before has a theory been developed which has given a key to the interpretation and calculation of such a heterogeneous group of phenomena...," he adds that, nevertheless, "the theory is apt to beguile us into error... because, in my belief, it is an incomplete representation of real things." After discussing how (in his opinion) quantum theory fails at the task of describing individual systems, he continues:

> But now I ask: Is there really any physicist who believes that we shall never get any inside view of these... single systems, in their structure and their causal connections, and this regardless of the fact that these singular happenings have been brought so close to us, thanks to the marvelous inventions of the Wilson [cloud] chamber and the Geiger counter [for detecting particles]? To believe this is logically possible, but it is so very contrary to my scientific instincts that I cannot forego the search for a more complete conception.

There was, of course, one physicist who believed precisely what Einstein could not.

Although I have stressed the conflict between Bohr and Einstein because of its intellectual importance, the two men were in fact lifelong friends and said many flattering things about each other's science. Yet Bohr is mentioned only once in Einstein's reprise of his decade-long battle with quantum mechanics, and that is in a reference to *the Bohr atom*. EPR is discussed, but not Bohr's response. It was not a chivalrous treatment of a vanquished rival.

After his summing up in 1949, Bohr continued the debate with Einstein, even after the dialogue became a monologue with Einstein's death in 1955. Bohr turned over the main points endlessly in his mind; whether rethinking the arguments or just savoring old victories one cannot know. Bohr died in 1962. On the blackboard in his office in Carlsberg Castle a single drawing, made the night before, could be seen. It was a sketch of Einstein's photon-in-a-box *Gedanken* experiment.

10.

The Post-War Heresies

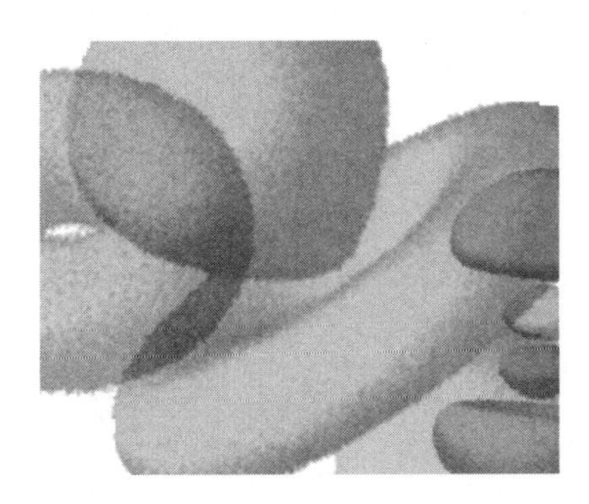

Einstein did not seem to know that this possibility
[a complete physical description of quantum phenomena]
had been disposed of with great rigour by J. von Neumann
. . . But in 1952 I saw the impossible done. It was in papers of David Bohm.

—John Bell, *Speakable and Unspeakable in Quantum Mechanics*

In 1945, Schrödinger was 58, Bohr was 60, and Einstein was past retirement age. Heisenberg, Pauli, Jordan, and Dirac were in their 40s. With the end of the war, another crop of bright young physicists would soon be emerging from the universities. Would they accept the revolutionaries' radical positivism—or follow Einstein and seek to restore realism in physics?

There had been several developments since the quantum revolution. In the 1930s, new particles, including the neutron and the first antiparticle, the positive electron or "positron," had been discovered (both in 1932). Dirac invented a version of Schrödinger's equation which incorporated relativity and electron "spin" and permitted him the triumph of predicting the positron's existence a year before it was found in cosmic rays.

In the first treatment of the creation and destruction of particles, Dirac computed the probability that an electron and a positron spontaneously annihilate each other, leaving behind two photons. Heisenberg and Pauli introduced a general theory called "quantum electrodynamics" of these processes, but to their dismay many quantities of interest came out infinite. The first particle accelerators, baby versions of today's behemoths, were constructed in England and America.

Shortly after the war, a group of young theorists—Richard Feynman, Freeman Dyson, Julian Schwinger, and Sin-itiro Tomonaga—created a new quantum electrodynamics that removed many of the infinities from the earlier theory and allowed them to compute some properties of elementary particles to an unprecedented degree of accuracy. Later, when modern computers became available, their theory generated numbers agreeing *in the first ten digits* with experiments. QED was a triumph for the reigning paradigm. (Yes, the accepted acronym is "QED"; physicists love to see the mathematicians cringe.)

One of the bright young post-war physicists who contributed to this breakthrough was Richard Feynman. Feynman was born in 1918 in New York City, educated at MIT and Princeton, and worked in the Theory Section of the Manhattan Project under Hans Bethe during the war. In 1948, while an assistant professor at Cornell, Feynman invented a novel approach to quantum mechanics he called the "space-time," or "path integral," method. Feynman first associated complex numbers called "path amplitudes" with classical particle trajectories. To compute the probability the particle arrives at some place, he added the amplitudes for all paths leading there, as one would probabilities for mutually exclusive events in classical probability. (The term "probability" must be used advisedly in this context, since these path quantities do not follow the normal rules for probabilities; see Chapter 16, "Paradoxes.") Then he computed the squared length of the sum as in Schrödinger's wave theory. Feynman used this technique to calculate some quantities in quantum electrodynamics in a way physicists found appealing, and many now regard it as the defining picture of the microworld. Feynman insisted "by particle I mean *particle*" and claimed his space-time picture banished waves entirely, but in his conception a particle could somehow sample all paths at once. Although Bohr objected vehemently to this partial resurrection of particle paths, Feynman's idea was only mildly heretical. (For more on Feynman's unique way of thinking, see Chapter 16, "Paradoxes," and Chapter 19, "Opinions.")

Another bright young man studying quantum mechanics in the early 1950s was David Bohm. Bohm was born in Wilkes-Barre in 1917 and attended the Universities of Pennsylvania and California. At the latter institution he was a student of Robert Oppenheimer. When Oppenheimer was picked as scientific chief of the Manhattan Project, the FBI, citing security reasons, refused Bohm permission to follow him to Los Alamos, so Bohm spent the war years working at the Radiation Laboratory in Berkeley. After finishing his doctorate, Bohm accepted an assistant professorship at Princeton.

At Princeton, Bohm adopted a time-honored strategy for understanding a puzzling subject: write a book about it. The textbook he completed in 1951 was the best account of quantum mechanics in its day and is still worth reading. (Bohm's text was somewhat unusual for its large ratio of words to equations. "Too much schmooze," remarked Eugene Wigner to Abner Shimony.) Among other accomplishments, Bohm gave a clearer account of the Copenhagen interpretation than any produced by Bohr and introduced a new version of the EPR experiment (now called the EPRB experiment), which soon replaced EPR's as standard in discussions. Two decades and several transformations later, it was this experiment that made its appearance on a laboratory bench.

For his experiment Bohm exploited atomic magnetism and the Stern-Gerlach effect described in Chapter 2, "Prologue, Part II: Quanta." (The experiment is sometimes described as using electrons, but the electron's net charge would probably make it unworkable.) He imagined a pair of magnetic atoms in a molecule which are separated in a source device and sent to pairs of Stern-Gerlach analyzers. Picture the source as at the center of a laboratory bench and the analyzers at each end; see Figures 12 and 13. Each analyzer contains an electromagnet and a pair of particle detectors. The atoms arriving at the analyzers are deflected either up or

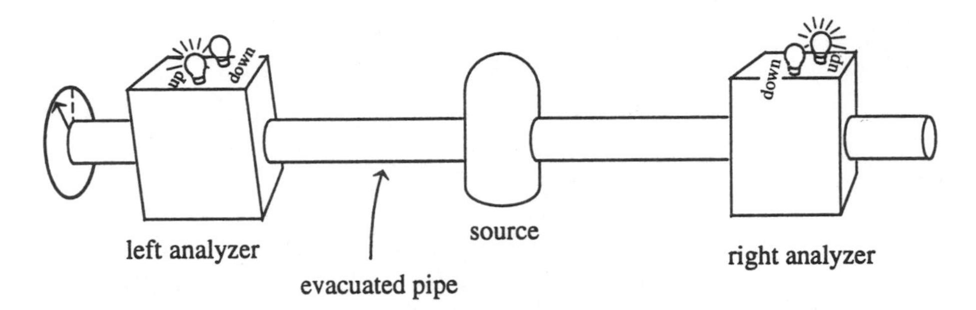

Figure 12

down by the magnet depending on the atom's "spin." Using U for "up" and D for "down," the output of the detector on the left over many runs of the experiment looks like this:

$$U \quad U \quad D \quad U \quad D \quad D \quad D \quad U \quad \cdots.$$

Remarkably, for a certain highly symmetric initial state of the two atoms the output of the detector on the right will be

$$D \quad D \quad U \quad D \quad U \quad U \quad U \quad D \quad \cdots.$$

That is, the output is exactly reversed from the left output, like the complementary strand of a DNA molecule. (See the Appendix for the derivation of this remarkable fact.) The EPR situation arises by rotating one or both

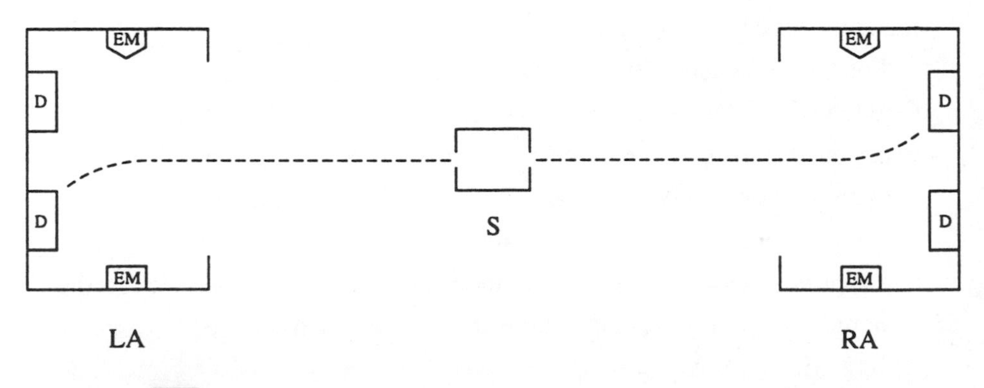

Figure 13

of the analyzers by 90 degrees around the axis passing through them. According to quantum mechanics, the values of an atom's "spin" in different directions do not have simultaneous reality, but obviously measurements on the right let you know either one on the left, so EPR would have argued that they are all real.

After completing his opus, Bohm's thinking underwent a radical shift. In 1952 he rejected Copenhagenism and, in two papers for *Physical Review*, advanced a different interpretation of quantum mechanics—one based, remarkably, on classical mechanics.

The starting point for this new interpretation was a pair of "physical" variables discovered earlier by Schrödinger. In Schrödinger's original theory, the state of a material particle had been described by a wave whose "height" at each point in space is a complex number. But it is meaningless to speak of a physical magnitude of, say, twice the square root of minus one. So Schrödinger introduced two other quantities that do have physical interpretations. These are, in the jargon of mathematical physics, a "scalar field" and a "vector field." (The usage is the same as in "poppy field" for a field of poppies.) The scalar, or numerical, quantity can represent the probability of finding the particle in a small volume centered at that point; the vector gives the rate and direction of flow of probability past that point. Ignoring their origin in quantum mechanics, these two quantities might as easily describe the depth and current velocity of a water wave. (Indeed, the picture is sometimes called "Madelung's fluid," after E. Madelung, who published it in 1926, although Schrödinger probably knew it earlier.)

Bohm next asked how this pair of physical quantities change in the course of time. A law governing their evolution follows from Schrödinger's for his wave. At this point Bohm saw a classical vision in his formulae: the Hamilton-Jacobi equation emerged. Thus, as frequently happens in scientific history, the concepts had passed through a complete circle: this had been Schrödinger's starting point a quarter-century before. With this insight a realistic interpretation became obvious. A particle might move on a classical trajectory and be subject to two kinds of force: the conventional force, due to gravity or any other force fields that might be present, and an unconventional "quantum" force, computed in a complicated way from Schrödinger's wave function. Thus, in Bohm's mind, the wave function acquired a real existence: it became simply a new force field joining those of classical physics—the electric, magnetic, and gravitational fields.

Bohm had rediscovered de Broglie's pilot wave in a mathematically more coherent form. After distributing his paper, senior colleagues informed Bohm of the earlier episode, and he added the remark that he could answer all of Pauli's earlier objections. Bohm had successfully added "hidden variables" to quantum mechanics without doing violence to its predictions. Ironically, considering Heisenberg's original motivations and von Neumann's book (which Bohm had read), the variable that was hidden was an old-fashioned, classical particle trajectory.

Bohm's model did not result in the restoration of realism in physics. His work was attacked on a variety of grounds, most overtly ideological. Léon Rosenfeld called Bohm's approach "a short-lived decay product of the mechanistic philosophy of the nineteenth century." Einstein had a less florid response, remarking in a letter to a colleague that Bohm's solution to the quantum puzzle seemed "too cheap." (Einstein was of course hoping for a revelation: a Unified Field Theory explaining both gravity and quantum phenomena.) Even de Broglie rejected it—although he later changed his mind.

Deciphering the effect of Bohm's unorthodox views on his subsequent career is problematic, for in the late 1940s the establishment was more interested in suppressing a different heresy. In 1949, Bohm was summoned before the House Un-American Activities Committee, where he refused to reveal whether he had ever been a Communist, or whether Communists had been active at the radiation lab in Berkeley. Although acquitted of contempt charges in federal court, Bohm was kicked out of Princeton and was subsequently unable to find a job anywhere in the United States. Bohm turned to Einstein for help and, armed with a letter of recommendation from the great scientist, found a position in Sao Paulo, Brazil (where he wrote his papers on quantum realism). The State Department then lifted his passport. Thus did the American establishment, after welcoming so many scientists fleeing persecution abroad, turn one of its brightest young physicists into a stateless refugee.

Bohm continued research in theoretical physics, making significant discoveries about the predictions of quantum mechanics. The "Aharanov-Bohm effect," the puzzling influence of a magnetic field confined to a cylinder on an electron which never penetrates inside, which he discovered with a student, Yakir Aharanov, in 1959, is probably his best-known contribution. Bohm moved to Israel and then to England, where he joined the faculty of Birkbeck College, of London University, in 1961. By all standards, he had a successful career.

In 1984, Bohm was invited back to Princeton to give the weekly physics colloquium—perhaps as a gesture of repentance, perhaps to see whether or not he had recanted his views. (On physics, that is.) Surprisingly, he consented. His talk, I recall, started out with some fireworks when a prominent Princeton theorist got into a public wrangle with the speaker which lasted for a quarter of his allotted time. (The point of contention, according to my memory, was the symmetry in quantum theory between position and momentum, which Bohm's theory violates by giving the particle's position special importance. Physicists, like most scientists, are infuriatingly attracted to debating these technical questions rather than getting down to the *real* issues.) If the hope was to score a definitive knockout, Bohm dodged the blow. After this initial excitement, the audience appeared to lose interest in the remainder of Bohm's talk.

Why does the physics community shun Bohm's realistic interpretation of quantum mechanics? Is it just rigid adherence to orthodox scripture? Although ideology undoubtedly played a role, it must be admitted that Bohm's theory was almost as mysterious as conventional quantum mechanics. In the first place, the "quantum force" acting on the electron was computed in part from the *probability law* of the particle's location. Since probabilities must be computed from averages over a set of experiments, it appeared that some kind of ghostly influence of the ensemble on a given one was implied. This is as plausible as suggesting that if an abnormally large number of people in my town misplace their wallets on a given day, I am more likely to lose my own. Second, the equations of

Bohm's model are unaesthetic when compared to Schrödinger's, although this may just reflect the simplifications the square root of minus one can make in equations. And since Bohm's equations are nonlinear, one would automatically appeal to Schrödinger's when attempting to solve them. Finally, one prediction, though not absurd, was hard to swallow. In the ground state of the hydrogen atom, the velocity of the electron around the proton equals zero, according to Bohm. The electron just sits there—albeit at a random position. (From a modern standpoint, a more serious objection to Bohm's theory is that it has not been successfully generalized to be compatible with relativity or to include particle creation and annihilation.)

These are serious objections—although they were not those of the establishment in 1952. On the positive side, the "pilot wave" resolves many paradoxes of quantum theory. As just one example, consider the two-slit *Gedanken* experiment from Chapter 6, "Complementarity," which Feynman claimed to be almost beyond rational understanding. ("... a phenomenon which is impossible, *absolutely* impossible, to explain in any classical way," Feynman wrote in his "red books" for beginning students, *The Feynman Lectures on Physics*.) Here is how it appears in the pilot-wave theory: when both slits are open, the wave passing through the slits undergoes interference and diffraction just as for sound or water waves. Although not apparent until revealed by a flux of particles, the pilot waves form a two-slit pattern on the screen. When particles come along, they follow the guiding wave and hence tend to appear at the maxima of the pattern and never at the minima. Now if one of the slits is blocked, after a time necessary for a disturbance in the wave to reach the screen the invisible pattern changes from two-slit to one-slit. If a particle then arrives, it behaves accordingly, for it always follows the counsel of its guiding wave. The paradox has vanished. (In a book on his "space-time" approach to quantum mechanics (Feynman and Hibbs, 1965), Feynman and co-author note that closing one slit *might* perturb the motion of the particle passing through the other slit, but dismiss this possibility as too implausible to consider.)

Bohm's model revealed this "paradox," von Neumann's "impossibility theorem," and the Berkeleian conceit expressed by some physicists that an electron is only there if you are looking at it to be ideological prejudices rather than logical consequences of quantum theory. As Bohm put it in 1952, "... the mere possibility of such an interpretation proves that it is not necessary for us to give up a precise, rational, and objective description of individual systems at a quantum level of accuracy."

Another aspect of Bohm's model is that it contains a peculiar form of "action at a distance." Far from being just another blemish in a flawed theory, this fact proved the key to a fundamental discovery. It is the subject of the next chapter.

In one of those coincidences that are by no means rare in science, another physicist constructed a realistic interpretation of quantum mechanics in the same year Bohm's work appeared. (One explanation for this coincidence may be that the quarter-century from the birth of quantum mechanics to the early 1950s represented the time required for the original arguments to die out or become so garbled as to lose their force. Another possibility is that a new generation absorbed the facts of quantum physics while escaping exposure to certain strong personalities, e.g., that of Bohr.) The Hungarian physicist Imre Fényes of the University of Debrecen also assumed that particles move on trajectories—but not the familiar trajectories of Newtonian mechanics. He was struck by the analogy between quantum mechanics and a branch of mathematical physics with roots in the 19th century: the theory of Brownian motion. The adjective "Brownian" refers, improbably, to observations made by an English botanist. I digress to discuss this celebrated episode and the remarkable developments in mathematics and physics it initiated.

In 1828, Robert Brown published in the *Philosophical Magazine of London* "A Brief Account of Microscopical Observations Made in the Months of June, July and August, 1827, on the Particles Contained in the Pollen of Plants," in which he described peculiar motions of small particles suspended in water. This motion, an irregular "dance," went on unceasingly. (Brown was not the first to observe this motion—it presumably had been seen by every microscopist since Leeuwenhoek—but Brown succeeded in elevating it from an irritation to an enigma.) The particles Brown observed included crushed rock, pollen grains, soot from smoggy London, even fragments chipped off the Sphinx. The natural first question was whether the grains were alive and, consequently, whether their motion could be attributed to the then-popular "vital force." Brown decided against the hypothesis, since the motion appeared to be universal. Nevertheless, he thought it probable that the particle's motion was caused by disturbances generated from within and not by the influence of the surrounding water.

In the second half of the century, there were numerous investigations of this ubiquitous phenomena, now called "Brownian motion." (Two other terms then in use were "titubation" and "pedesis." The first survives as a term used to describe the staggering gait of patients with certain nervous disorders; the second has disappeared—at least from my dictionary.) Brown's explanation soon lost favor, but other explanations invoking hydrodynamical or electrical forces fared no better. The phenomena remained an enigma until Einstein took it up in the year 1905—the same year he published the theory of relativity.

Einstein was trying to find evidence for the reality of atoms—an understandable goal given the skepticism of Mach and his followers, which persisted well into this century. Einstein reasoned that the key to proving that atoms exist lay in *fluctuations*: random changes that would be inexplicable to the pure thermodynamicist or "energeticist." If atoms are real, there should be fluctuations in the force on a pollen grain immersed in a water bath, due to statistical variations in the number of bombarding water molecules. Although the impact of an individual atom does not have enough "kick" to produce an observable effect, the random variation in the number of impacts, Einstein observed, should cause the grain to wander aimlessly about, like an inebriated bacterium. Einstein showed that the particle should make increments in distance proportional to the *square root of time* and related the ratio

$$\frac{\text{distance moved}}{\sqrt{\text{time elapsed}}}$$

to the temperature of the water and Avogadro's number—permitting a determination of the latter using a microscope, some water and pollen, a stopwatch, a ruler, and a thermometer.

Unfortunately, at the time Einstein was not engaged in experiments in statistical physics but in investigating patent claims, so he produced no data to support his calculations. Later, the French experimentalist Jean Baptiste Perrin filled in the required information, yielding a value for Avogadro's number of 6.0×10^{23}, close to the modern value. (Perrin received the Nobel Prize in 1926 for carrying out these measurements on Brownian particles.) Remarkably, in 1905, Einstein found *three* independent methods for computing Avogadro's number; Planck's quantum theory provided yet another, and they all agreed. Atoms, everyone concluded, were real.

The kind of motion postulated by Einstein was also very interesting to mathematicians. It seemed to be a continuous version of the "drunkard's walk" of probability theory, in which a drunk is assumed to take discrete steps completely at random. Was there a random process *with continuous paths* that mirrored the behavior of the drunkard's walk? In 1923, the American mathematician Norbert Wiener proved that such a process exists. The paths of this process are continuous (no instantaneous jumps) but otherwise are unlike Newtonian paths. They are so "crinkly" that it is impossible to define a velocity at any point of the path, or to measure the length in the ordinary way (one gets infinity if one tries). Furthermore, the paths occupy space in such a strange manner that they can be said to have a "dimension" greater than one. (Curves with this property are now called "fractals" since they may have a dimension that is a fraction rather than an integer.) Since Wiener's original paper, the study of the properties of Brownian paths has become a cottage industry in mathematics, with scores of abstruse papers published annually—a strange legacy for a botanist.

My own research involved the study of generalizations of Brownian motion, and I thought I knew the history fairly well. So I was startled when a friend told me one day over coffee that the theory of Brownian motion had been invented five years before Einstein by a man studying the French government bond market!

"Ach... what was that you said?" I sputtered, dabbing at the the espresso I had just dumped in my lap. My friend agreed to furnish the reference, and the next day I was in possession of the thesis of one Louis Bachelier, entitled "Theory of Speculation," which had been accepted by the Faculty of Sciences of the Academy of Paris on March 29, 1900. In it Bachelier developed the theory of price fluctuations in a market, answering the question, "Does the recent history of a stock or bond give any information about its future performance?" with a resounding *No*. The price of a stock, Bachelier asserted, could be understood as performing a kind of continuous random walk: like the drunk in the standard metaphor, it completely forgets where it was going at each instant. His theory had interesting consequences concerning such tools of speculators as "puts," "calls," "contangoes," "futures," "call-o'-mores," and other devices completely foreign to a mathematician who stupidly deposits his check in the bank each month, letting others have all the fun.

Bachelier knew the basic equation for increments of a Brownian particle ("price increments" for Bachelier), the square-root law, and many other facts. His results were not completely rigorous—not surprising considering the date—but they were coherently developed and clearly explained. Mathematicians sometimes make snide remarks about economists and similar types, whispering that certain persons have won Nobel Prizes for rediscovering calculus lemmas. How appropriate that a student of finance should have invented half the formulas describing Brownian motion, years before the physicists and mathematicians who normally get the credit. As Paul H. Cootner of MIT put it,

> So outstanding is his work that we can say that the study of speculative prices had its moment of glory at its moment of conception.... It also marked the beginning of the theory of stochastic processes, a beginning that went unrecognized for decades.

To salvage their pride, mathematicians might take refuge in the fact that Bachelier's thesis advisor was Henri Poincaré, one of the most celebrated mathematicians of all time, who *lectured on Brownian motion* in 1900 at the International Congress of Physics in Paris. Perhaps mathematicians should credit Poincaré with inventing the theory of Brownian motion, thus offending the physicists as well as the economists.

Returning to physics and to the year 1952, we find Imre Fényes contemplating Schrödinger's equation written out in the "physical" variables at the same time as Bohm. Like Bohm, Féynes sees particle paths beneath these ugly formulas, but paths that are intrinsically random, like Brownian paths. Consequently, the classical concept of velocity disappeared in Fényes' model. Féynes apparently was unaware of Wiener's mathematical theory, so he simply assumed that his processes made mathematical sense. He was aware, however, that "velocity" lost its classical meaning. This was a radical new hypothesis: Unlike Bohm, Fényes did not try to restore the classical picture in particle physics.

Fényes theory, like Bohm's, was not without difficulties. At the end of his highly original paper, Féynes gave a derivation of quantum mechanics from statistical assumptions and a least-action principle, a result that should have generated a stir. But, unfortunately, Féynes expression for the energy of his diffusing particle contained a peculiar contribution (which he called the "pressure energy") computed from a knowledge of the probability law of the particle. This left his theory open to the same objections as to Bohm's.

Like Bohm's, Fényes' work exploded von Neumann's claim to have ruled out realistic models. Fényes could not resist a jibe at his famous fellow countryman:

> If we consider... the von Neumann proof of the impossibility of the existence of "hidden parameters," we may perhaps be inclined to think that any attempt to insert quantum mechanics into the framework of "classical" statistics is a design as vain and as doomed from the beginning to shipwreck as e.g., the attempt to derive a "perpetuum mobile."

To the contrary, writes Fényes,

> ... we will see that every individual characteristic of quantum mechanics which differentiates it from classical physics has its origin in the statistical mode of thought and beyond this has no differentiating features.

Quantum mechanics, says Fényes, requires no embrace of "uncertainty," no struggle to think in complementary pictures. All one need do is adopt a statistical approach—and this is no more radical a step for physics than it was for actuarial science.

The reaction of the physics establishment to Féynes' paper was the same as to Bohm's: a few misdirected attacks, then silence. I have no information on Fényes' subsequent career.

Just as Bohm's classical interpretation of quantum mechanics was not completely without precedent, so too Fényes' discovery of a connection between quantum mechanics and Brownian motion—although Fényes was the first to take the idea seriously. In Richard Feynman's 1948 paper on his "space-time" formulation of quantum mechanics, he remarks in an aside that the paths of importance in his approach are "of a type familiar from the study of Brownian motion." Bohm also noted this connection. He remarked, also in passing, that the "quantum force" in his model was in many cases (especially during measurements) so irregular that the particle's paths would resemble those of Brownian motion. Wiener himself attempted to relate his random process to physicist's theories, developing, together with Armand Siegel, another "hidden variable" model in 1953, shortly after Bohm and Fényes.

In 1966, Edward Nelson, a Princeton mathematician, startled the unsuspecting audience in a probability course by announcing a derivation of quantum mechanics from classical mechanics and Brownian motion. (I remember being equally startled when I read the book that resulted from this course, innocently titled *Dynamical Theories of Brownian Motion*, a decade later.) Nelson had rediscovered Fényes' basic idea and rendered it more plausible. (Although Bohm's model had acquired some notoriety, Fényes' was completely obscure; I believe Nelson when he says he was totally unaware of Féynes.) Nelson's derivation involved a stochastic substitute for Newton's second law of motion (force equals mass times acceleration); this required a redefinition of "acceleration," since this concept made no more sense for Brownian paths than did "velocity." Nelson's choice had a slightly ad hoc character but was less objectionable than Féynes' "pressure energy." In a sense, Nelson showed that quantum mechanics could be derived from the principle that elementary particles are subject to a universal jiggling of unspecified cause, together with the principles of mechanics, suitably modified. By contrast, the "derivation" of quantum mechanics given in most physics texts replaces the intelligible (position or momentum) by the inexplicable (infinite-dimensional matrices) with the flippant air of a conjurer turning a scarf into a songbird.

Since the middle 1960s, Nelson and his students and co-workers have developed "stochastic mechanics," as the Brownian approach is usually called, to the point where it covers the same ground as nonrelativistic quantum mechanics (including phenomena such as electron "spin"). Stochastic mechanics has, in my judgment, fewer objectionable features than Bohm's model, and it would have been wonderful to have a derivation of quantum mechanics from more intelligible principles. (Bohm's reasoning begins with Schrödinger's wave as a given.) But life is rarely so simple. In 1989, after the first draft of this chapter—with a somewhat different ending—had been completed, an ex-student of Nelson's named Timothy Wallstrom sent me a letter and unpublished manuscript. Wallstrom said he had discovered certain "anomalies" in stochastic mechanics. This was an understatement: Wallstrom had found a striking discrepancy between quantum mechanics and stochastic mechanics. For many cases, *including the hydrogen atom*, they do not give the same predictions.

Wallstrom's observation is topological in nature and helps in understanding the status of the heretical theories in which particles take paths. I postpone further discussion until Chapter 18, "Principles," and only remark here that Wallstom's discovery suggests these theories are too far on the "particle side" of the wave-particle dichotomy to be entirely correct.

Afterword. Imre Fényes died sometime in the 1970s. David Bohm died in 1992.

11.

Bell's Theorem

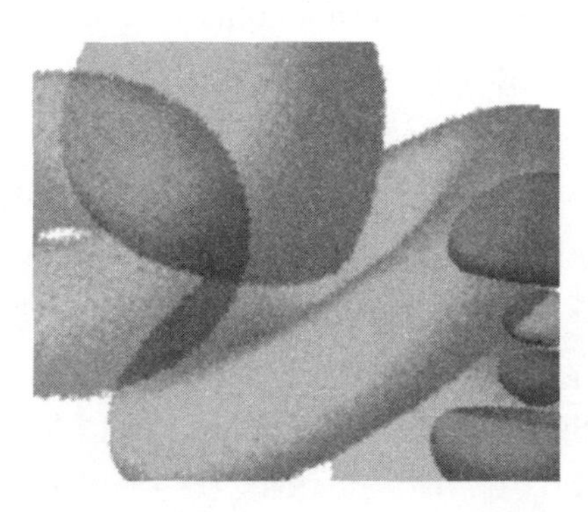

In the early 1950s, a physics student in Northern Ireland of unusual critical ability noticed a puzzling situation. Einstein, the student knew, had claimed quantum mechanics gave an inadequate account of atoms, and he had read about von Neumann's "impossibility proof" in a popular book. But then was not one of these geniuses wrong?

John Bell was born in Belfast in 1928 to working-class Protestant parents. His father was a horse trader who quit school at eight, and others in his family were blacksmiths and farm workers. Bell, a bookish child, barely escaped a life of manual labor himself; his siblings left school for work at 14, but luckily some money was raised to continue his schooling. (Universal education had not been introduced in the United Kingdom.) Bell graduated from Belfast Technical High School at age 16, the first in his family to complete a secondary education.

Too young to start college, Bell found a temporary job as a lab assistant in the physics department of Queens College, Belfast. There, surrounded by coils and magnets and voltmeters, he chose his future profession. Bell graduated in 1949 and acquired a Ph.D. in 1955 from the University of Birmingham in England. His thesis advisor was Sir Rudolf Peierls, a leading figure in statistical mechanics and quantum theory, who was the first to prove that a system of spins could spontaneously form a magnet. After his doctorate Bell worked for a few years for the British atomic energy agency at Harwell before joining the European high-energy physics laboratory (CERN) in Geneva in 1959. At CERN he participated in the design of Europe's first particle accelerator and married another accelerator designer. On a visa application he once described himself as a "quantum engineer."

While still a student Bell discovered von Neumann's mistake and something even more interesting in Bohm's papers. But following these leads presented obvious dangers to his career. When Peierls asked him to choose between speaking at a seminar on the foundations of quantum mechanics or on accelerators, he chose accelerators. He wrote no skeptical papers about quantum mechanics. "I walked away from the problem," he remarked many years later. A decade passed. Then in 1963, on sabbatical at Stanford, Bell returned to his original interest—and wrote two spectacular papers.

In the first paper Bell refuted the conclusion of every "impossibility theorem" proven to that date, including von Neumann's. (The paper appeared in 1966 due to an editor's mistake in scheduling.) Then he proved such a theorem himself—one that has survived unscathed to this day. Bell had noticed that in Bohm's model for two or more particles there was a striking *nonlocality*, or form of action at a distance. The nonlocality was lurking in Bohm's expression for the "quantum force." In Maxwell's theory of electromagnetism, a particle could only be influenced by force fields in its immediate vicinity. Relativity, through its prohibition of influences propagating faster than light, ruled out anything else. But a glance at Bohm's equations for a pair of particles revealed that *both* could be affected by a magnetic field present at the location of *one of them*.

Bohm knew about this nonlocal influence in 1952 but dismissed the issue as unimportant. After all, he remarked, the forces involved were tiny. To exploit them to send a signal faster than light, one would have to measure a particle's position with unheard of precision. And so forth. (A cheerful disregard for minor anomalies can be a healthy attitude in a creative individual. If every theory were required at the outset to be free from glitches, few would get published.)

At first Bell tried to find a model like Bohm's but without the troubling nonlocality. But nothing worked. "Then I constructed an impossibility proof," he later remarked, unhelpfully. The resulting paper, which appeared in 1964, has become one of the most famous in 20th-century physics. It is five pages long; the crucial section, entitled simply "contradiction," covers two pages. Although there are about a dozen equations on those two pages, the argument is surprisingly simple.

Bell derived his argument from some careful cogitation on Bohm's version of the EPR experiment. Recall that the analyzers in the EPRB setup are located at the ends of a laboratory bench, with a source of particles in the center. The analyzers can be rotated around the axis passing through them and the source by an arbitrary angle. Each has two particle detectors, one arbitrarily labeled "up," the other "down." Recall also two facts about the way the particles behave in the analyzers at the ends of the bench:

(i) Watching only one analyzer, we see a completely random sequence of outcomes—as though the particles flip a fair coin to decide where to go.

(ii) Yet if the analyzers are aligned, the detections are always complementary: up on one side means down on the other side.

One other fact was known: if the analyzers are oriented perpendicularly to each other, no correlation is evident in the data. These facts had been verified in experiments. (See Chapter 13, "Testing Bell.")

Bell decided there was more juice to be squeezed out of this experiment, by considering angles *other* than zero and ninety degrees. For instance, if we choose three angles spaced by 120° (say 0°, 120°, and 240°), the data from the experiment are the following. From averages over many runs, we compute frequencies of the four possible outcomes (up-up, up-down, down-up, and down-down) with analyzers rotated to the various angles selected. Since these frequencies depend only on the relative angle, we need only consider one case, say the left one at 0° and the right one at 120°:

Outcomes:	U, U	U, D	D, U	D, D
Frequencies:	3/8	1/8	1/8	3/8

Thus one particle goes up and the other goes down in one run out of eight on average, while they both go up three-eighths of the time. These are the frequencies Bohm had derived from quantum mechanics for his experiment. See the Appendix for the derivation of these numbers. Bell proved that these innocent figures imply a strange and surprising fact.

Bell's conclusion becomes even more striking if put in a mundane context. Imagine two slot machines, identical in appearance, located at gambling casinos in different cities. On either machine you can play one of three games. After putting in your quarter and choosing your game, the machine—perhaps after a lot of blinking lights and whirring sounds—responds "You're a Winner!" and returns a payoff, or "Sorry, You Lose!" and keeps your money. (I am aware that this game might not prove very popular in Las Vegas.) There is an interesting and possibly lucrative experiment to carry out with these two devices. Using cellular telephones, a pair of gamblers could arrange to simultaneously play the games in various combinations—I play game one, you play two, and so forth—and test them for correlations. That is, determine if they have a tendency to produce wins simultaneously, or if there is some other relation between the patterns of wins and losses. For instance, if they operate by a list of predetermined outcomes built in at the factory, one could make a bundle by discovering this fact. If the "anti-correlations" of the EPRB experiment were present (with a small lag time to permit transmitting "play game II!" when somebody has just lost on that game), you might also turn a profit.

I invite the reader who likes a mathematical challenge to prove, before I present Bell's argument, that the winning frequencies listed above could never occur in this situation, no matter how the machines are constructed.

I next present Bell's theorem, with some pictorial and historical amplifications. Bell was seeking a contradiction between realism, relativity, and quantum mechanics; so, naturally, he began by assuming all three. The restrictions of relativity and the predictions of quantum mechanics are generally agreed upon, but as usual realism requires a special definition. An appropriate hypothesis is that there is a "complete set of hidden variables" that determines the outcome (which way the particles go) on each run. These variables might describe things located in the source, in the analyzers, or carried by the particles among them. The magnetic fields in the analyzers might have a complicated microscopic structure, which direct the particles the way arteries direct blood cells. Or the particles might come in various colors or sizes or shapes, and this determines where they go. They might even carry little messages to be "read" by the analyzers, like the list of 44 countries inscribed on a grain of rice that one reads with a microscope. In short, the realism postulate is that such variables exist—but their precise nature is irrelevant.

For simplicity, and as an aid to visualization, I represent "the hidden variable" (it is simplest to think of it in the singular) as a point in a square of side one. This also makes it easy to incorporate randomness in the picture: we can imagine dropping a grain of sand or a thumbtack on the square. Areas then equal probabilities. But this choice is of no consequence for the argument, since Bell's reasoning requires only elementary logic, together with the rules for combining probabilities (which are the same as the rules for combining areas: if several subregions add up to a given region and do not overlap, just add their separate probabilities to get the total).

Consider the event "a particle heads up at the left analyzer." Draw three closed curves in our unit square, each enclosing the hidden variables that yield this outcome for one of the chosen angles of the left analyzer. In Figure 14, I have drawn these three regions, labeled *I*, *II*, and *III*, as overlapping circles. Now consider that part of circle *I* that does not lie in circle *II*. This is the set of hidden variables for which the particle heads "up" on the left if that analyzer is oriented at the first angle and "down" if oriented at the second. What probability should this region have? If we can only observe the left analyzer, we could not say anything about it, since we cannot give this analyzer two orientations at once.

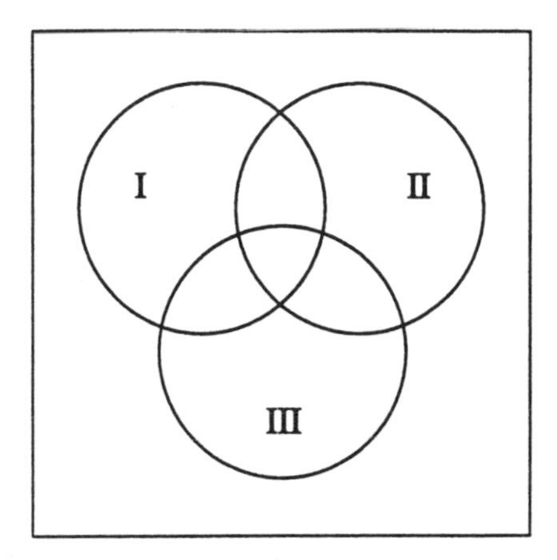

Figure 14

But the presence of the right analyzer and the "anticorrelation" provides the desired information (as Bohm pointed out): we need only note the frequency of the event "left analyzer: particle goes up, right analyzer: also up," with the analyzers at the selected angles.

Now the idea is to derive some interesting conclusion about the areas (probabilities) of these three regions. Here is a simple fact that the reader will readily prove:

The area of the part of circle *I* not in circle *II*

—plus—

the area of the part of circle *II* not in circle *III*

—plus—

the area of the part of circle *III* not in circle *I*

. . . is at most 1.

This is a simple exercise in plane geometry. The proof is given at the chapter's end.

For those who prefer probability to geometry, I state the same fact another way. (See the Appendix for the definitions of "event" and "probability".) Using the notation *P*[ ] for "the probability of the event enclosed in brackets is . . . ," and referring to events at both analyzers, the inequality becomes

$$P[\text{up at angle } I \text{ on left, up at angle } II \text{ on right}]$$
$$+ P[\text{up at angle } II \text{ on left, up at angle } III \text{ on right}]$$
$$+ P[\text{up at angle } III \text{ on left, up at angle } I \text{ on right}] \leq 1.$$

Either of these two statements is called "Bell's theorem" or "Bell's inequality." (Bell's was different but equivalent to ours; see the Notes.)

Plugging the numbers from quantum mechanics into either statement, we see that it asserts

$$3/8 + 3/8 + 3/8 = 9/8 \quad \text{is not greater than 1.}$$

We have a paradox. Our logic has been impeccable, but the conclusion is absurd. Might there be some dubious assumption in the argument, similar to von Neumann's postulate about adding averages? But what is most striking about Bell's argument is that it seems to postulate little more than the existence of a real world. Luckily for the realist, there is at least one hidden assumption: no action at a distance.

"Action at a distance" may sound like some bizarre parapsychological notion, but it actually first appeared in Western science in Newton's theory of gravity (1687). Newton's gravitational law reads, in modern language, "The gravitational force between two (spherical) objects is inversely proportional to the square of the distance between them, and directed *at each instant* along the line connecting their centers." Thus, in Newton's conception, one could send a message instantaneously to Jupiter by vibrating a large mass on Earth and detecting the resulting wiggle in a finely suspended needle on Jupiter. This is action at a distance (in its least subtle form).

In Newton's time, the followers of René Descartes lodged strong protests against this idea as deviating from the philosophically proper mechanical view of Nature into a blatant mysticism. (Newton, a devout theist, was not troubled by this accusation; he thought his conception left room for God to order his creation.) In the 19th century, Maxwell's theory of electricity and magnetism did not rely on action at a distance, nor did the "continuum" theories of elasticity or hydrodynamics. Newtonian gravity became the anomalous case among physicists' theories. After a decade-long struggle, Einstein succeeded in banishing action at a distance from physics with his general theory of relativity (1915). With the success of Einstein's program, most physicists thought action at a distance had been permanently eliminated, to the betterment of science.

Action at a distance, Bell noted, provides an escape from his theorem. His argument implicitly assumed that fiddling with the apparatus on the right could not instantaneously change the hidden variables affecting what happens on the left—say by moving circle *I* around on the figure. (An explicit example will be described in Chapter 20, "Speculations.") The analogue in the gambling scenario would be the discovery that pushing game button *II* on one slot machine sends an electronic message to the other one, which moreover changes its workings. Since the "ends of the bench" in the EPRB experiment might be separated in principle by any distance—light years, say—such instantaneous signaling would violate relativity. Aware of this assumption, Bell stated his conclusion very carefully:

> In a theory in which parameters are added to quantum mechanics to determine the results of individual measurements, without changing the statistical predictions, there must be a mechanism whereby the setting of one measuring device can influence the reading of another instrument, however remote. Moreover, the signal involved must propagate instantaneously, so that such a theory could not be... [compatible with relativity].

Abandoning relativity is, for most physicists, unpalatable in the extreme. With its long list of astonishing predictions—motion causes measuring rods to shrink, clocks to slow down, and mass to increase; $E = Mc^2$; and much more—all well verified by experiments, Einstein's 1905 special theory of relativity is as well established as any theory can be. Furthermore, Minkowski's "spacetime," as generalized by Einstein, is the stage on which all physics takes place (more discussion in the next chapter). To propose abandoning relativity would be like asking, if the physical world were a play, for a new theater rather than a new script or actors.

The dilemma Bell established in 1964 is severe but not entirely a matter of black or white. Although Bell's argument was completely rigorous, when real experiments were investigated that might verify his conclusions, yet another hidden assumption was identified. But this was realized only a decade later, after the crucial statement in Bell's summation had been clarified: what, precisely, is meant by "an influence on the reading" of a distant instrument?

Proof of the theorem: Note that the three regions (events) described have no overlap (are mutually exclusive), so the sum of their areas (probabilities) must be less than 1.

Afterword. John Stewart Bell died unexpectedly in October 1990 of a cerebral hemorrhage. He was 62 years old. Although Bell was best known for the theorem presented above, which ended hopes for an Einsteinian "field theory"–type solution to the quantum mystery, realists everywhere recognized Bell as one of their own. This they knew from his writings about quantum mechanics over the years, which represented a ray of clarity and objectivity amidst a jungle of vagueness and ideology. As well as being a productive scientist, Bell was a practitioner of a craft not thought admirable or even legitimate by most scientists: he was a *critic*, and a brilliant one at that. He was an Edmund Wilson or a George Orwell of physical science.

12.

Dice Games and Conspiracies

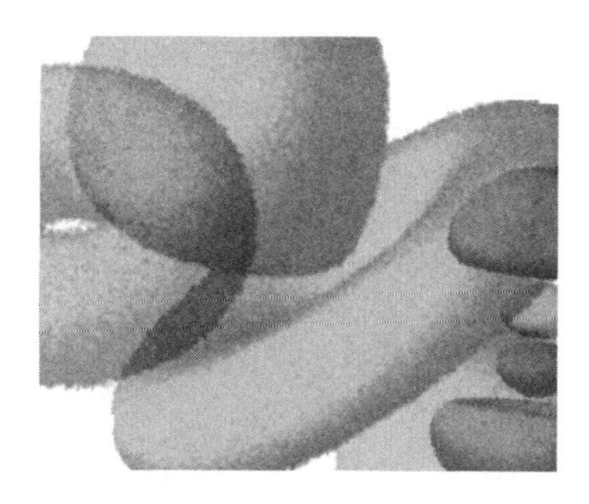

Ever since Newton, two forms of influence or causation have vied for supremacy in physicist's theories. Descartes and his followers thought local or contiguous action—material particles colliding and rebounding, for example—the essence of mechanical explanation; Newton evoked their wrath by advancing a form of action at a distance. In the 19th and 20th centuries, Maxwell and Einstein restored the local picture in their continuum theories of electromagnetism and gravity, and the cycle seemed to have come to an end—until Bell proved his theorem in 1964.

The modern physicist's understanding of causation derives from an insight of one of Einstein's teachers, Hermann Minkowski. Minkowski taught geometry and advanced mathematics at the Eidgenossische Technische Hochschule in Zürich where Einstein studied (and was later Professor), but Einstein paid little attention. (Einstein poked about in the laboratory while his friend M. Besso attended class and took notes. Einstein later regretted his failure to acquire a sound mathematical education.) Hence it was only fitting that Minkowski was the first to realize that Einstein's mathematics implied a unification of space and time. The reason lies in the transformation laws relating the space and time measurements of observers in different states of motion. In pre-Einsteinian physics, these rules kept the roles of space and time separated; but in relativity theory they became intertwined in an almost symmetrical way. This compelled the conclusion that time, like space, is a passive *dimension* of reality rather than an active agent. Minkowski introduced a four-dimensional continuum encompassing the spatial extent of the world together with its past, present, and future, and now called "space-time." As Minkowski put it in 1908,

> Henceforth space by itself, and time by itself, are doomed to fade away into mere shadows, and only a kind of union of the two will preserve an independent reality.

For a few years Einstein resisted Minkowski's proposal, perhaps because of its abstract mathematical character. (Einstein was a *physicist*, it should be remembered, not a mathematician.) Then he capitulated and, after educating himself in the non-Euclidean geometry of the 19th-century German mathematician Bernard Riemann, identified the "warping" of space-time with gravitation in his general theory of relativity (1915). Sadly, Minkowski died in 1909 without having witnessed the blossoming of his ideas in the work of his ex-student, but his contribution is memorialized in the accepted term for the un-warped space-time of special relativity: "Minkowski space."

In discussions of causation physicists often use a graphical device called the "space-time diagram," an attempt to represent Minkowski's four-dimensional continuum on a two-dimensional sheet of paper. In a space-time diagram a vertical axis represents the temporal dimension, and a single horizontal axis represents the three spatial dimensions; see Figure 15(a). Familiar objects take on an unusual appearance in a space-time diagram: they extend in an extra direction. For instance, a particle, represented simply by a point on a conventional drawing, becomes a curve, called the "world-line" of the particle. This curve is made up of all the events at which the particle was present: it is the whole "life" of the particle, from its creation to its annihilation. With the convention that the speed of light is equal to 1, world-lines of particles traveling at the speed of light, photons or neutrinos or other "massless" particles, become lines inclined 45 degrees to the vertical. Massive particles, which must travel

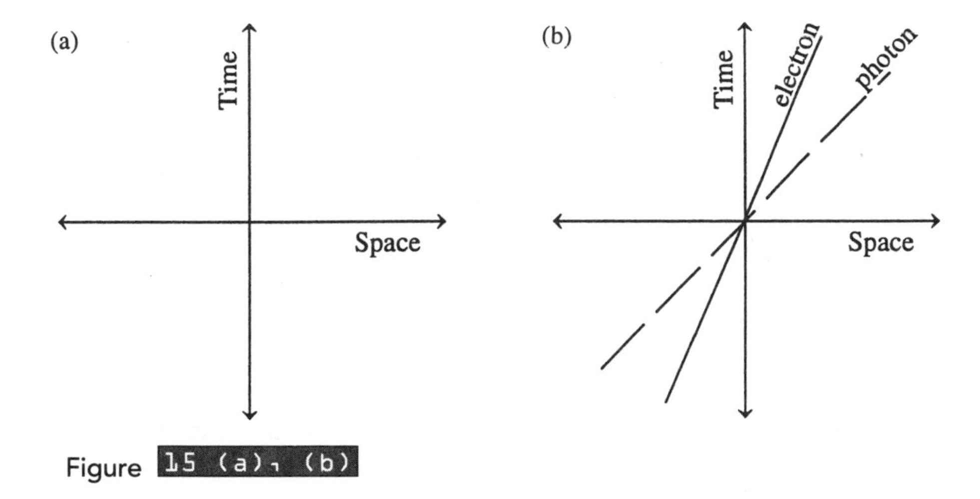

Figure 15 (a), (b)

at speeds less than that of light, have world-lines with greater slopes; see Figure 15(b).

Where is the "present" in such a diagram? That depends literally on your point of view. A "stationary" observer defines a vertical world-line and a family of horizontal lines, with each horizontal line representing "space at a fixed time" for that observer. But another observer in motion relative to the "stationary" observer will slice up space-time in a different way; see Figure 15(c). This was one of the reasons Minkowski concluded that space or time individually has no meaning. One has a tendency to imagine the present moment moving upward in these diagrams and even

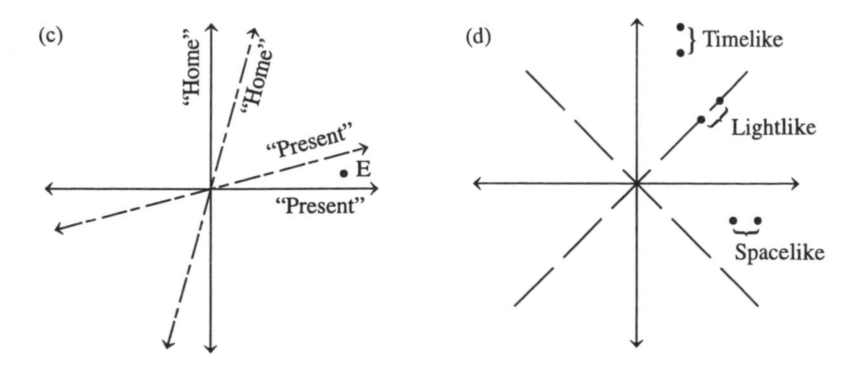

Figure 15 (c), (d)

to think of "me" as moving up my world-line, like a monkey climbing a vine. But this reflects a failure to adopt the relativistic world view. However much I may believe in this illusory journey from birth to death, in relativity theory I simply *am* my world-line, and hence I exist as an immutable fact of the universe—a thought some people have found comforting.

The points making up the space-time continuum are called, naturally, "events." Events can be related in one of three ways: they can be "timelike," "lightlike," or "spacelike" related. If an observer's world-line can pass through the events, they are timelike related. (Imagine the observer carrying a clock.) If the world-line of a photon can connect the events, they are lightlike related. Otherwise, they are spacelike related. (See Figure 15(d).) Note that two events are timelike related if for some observer they are located at the same place, while they are spacelike related if for some observer they occur at the same instant. But all observers agree on the kind of relation that exists between a pair of events, since they all agree, as Einstein proved, on the velocity of light.

The space-time diagram makes it simple to understand causality in Einstein's world. In Figure 16(a), an event "*a*" is labeled, and a region called the "backward light-cone" of *a* is shaded. This consists of all the events in the past of *a* either lightlike or timelike related to *a*. This region is also referred to as the "causal past of *a*." In Figure 16(b), I show the causal past of a whole region *A*, made up of all the causal pasts

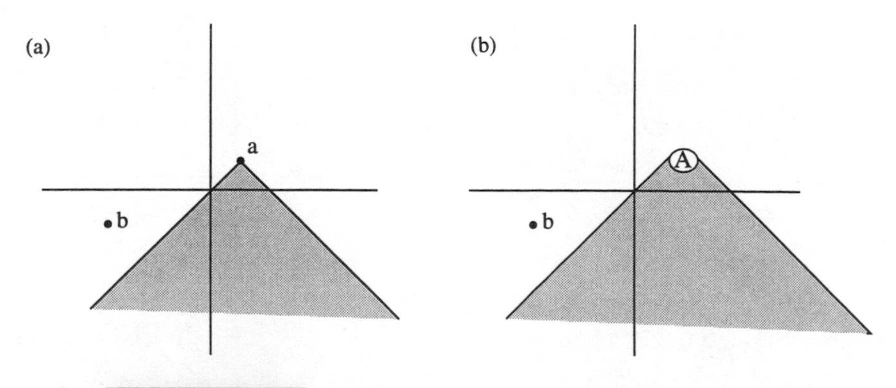

Figure 16 (a), (b)

of events in A. This terminology has been chosen to reflect a crucial implication of relativity: if an event b is to be the seat of a cause whose effect is manifested at a, then b must be in the causal past of a. There are two entirely convincing reasons for this. The first is simply logical consistency. If b is spacelike related to a, then for some observer they are simultaneous. To this observer a cause-effect relation would represent instantaneous signaling between distant points. (Provided it is possible to build a "transmitter" producing the desired cause at b—a point not without interest!) But Einstein's starting point in 1905 was the impossibility of such signaling; if we allow it, why have we forsaken Newton for Einstein?

The second reason concerns spacelike separated events: their time-order has no absolute meaning. Two observers moving relative to each other may not agree on which event came first. Therefore, if b is space-like related to a, some observer may find that an effect preceded its cause, involving us with the celebrated Grandfather Paradox beloved of the science-fiction writers: man travels back in time *via* Wellsian time-machine; man kills grandfather before grandfather can reproduce; man becomes troubled over the logical impossibility of his own existence; reader throws book away in disgust, resolving to seek out authors not prone to committing a logical *faux pas* for plot's sake. (My apologies to the ghost of H.G.)

I pause briefly to continue the explanation begun in the last chapter of why most physicists do not expect any future discovery to overturn special relativity. I first note that Einstein's general theory of 1915 might be said to contradict special relativity, but this is an overstatement. In general relativity the light-cones may not all be aligned with each other, due to the warping of space-time, but the finiteness of the speed of light and the basic trichotomy of causal relations remain unchanged. Einstein's choice of terminology indicates the true relationship of the two theories.

One might, however, imagine another material realm in which waves or particles propagate faster than light. Certain physicists prone to undisciplined theoretical and etymological speculation have dubbed this imaginary stuff "tachyonic matter" (from the Greek *tachys*, meaning "quick"); particles traveling faster than light are supposed to be called "tachyons."

These theorists postulate superluminary motions while maintaining their faith in Maxwell's and Einstein's theory—a remarkable mental feat. Besides the fact that tachyons would have imaginary mass—this is worse than for Schrödinger's wave; mass is *mass*, not some intangible probability amplitude—the higher signal velocity permits cause-effect sequences that are impossible according to conventional theory.

One can even set up a quite ridiculous case: two cars speed away from each other along a road. The driver in the first car pushes a button on a tachyon emitter, sending a tachyon in the direction of the second car. It arrives at that car *earlier* than it was sent and is detected. Instantly, the driver in that car triggers her own tachyon emitter. A moment *before*, the tachyon arrives back at the first car and is detected. Unfortunately for the first driver (and for our logic), his tachyon detector is wired up to a detonator and a stick of dynamite, which blows him to bits before he can push the original button. See Figure 17. (The Notes reveal how to set up this ridiculous scenario.)

One also hears from time to time from scientists who claim a direct shoot-down of relativity. Some are mathematicians or engineers (more rarely, physicists) who never quite convinced themselves that relativity is internally consistent. Reading Einstein's original papers is the antidote I prescribe for anyone so afflicted. Others are astronomers who divide the distance some celestial object apparently moved through space over a year's time by the number of seconds in a year and get a value greater than 3×10^8 meters per second—the speed of light. Ignoring the consistency of modern physics, the functioning of particle accelerators (which incorporate relativity effects such as the increase of mass with velocity in their design), and the Copernican principle (identifying celestial and

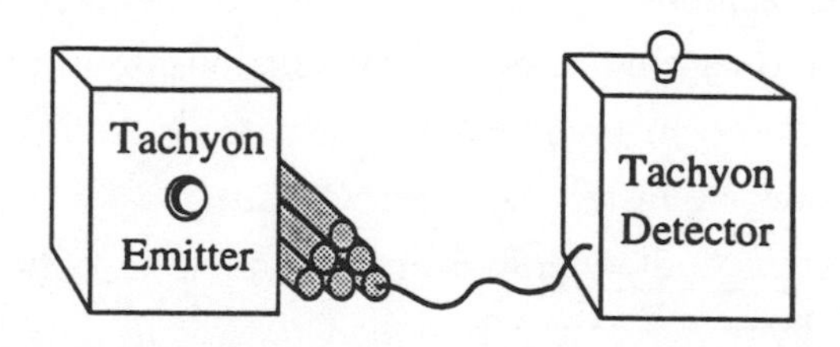

Figure 17

terrestrial physics), they are then prone to declaring the death of relativity. It invariably turns out that the estimate of the distance to the object was uncertain by a small amount, say by a factor of a thousand.

In 1964, Bell left dangling the issue of what kind of nonlocality must exist if quantum mechanics is correct. Beginning in 1971, he and some American physicists interested in turning EPRB into a real experiment took up again that impossible game (played earlier by von Neumann) called "list your assumptions!" In a realistic theory of atomic physics, what assumption would ensure "locality," meaning causal relations consistent with Einsteinian relativity? This question is complicated by randomness in a way I now explain.

Suppose I wish to send the message "buy" or "sell" to my stockbroker in New York from my house in Seattle. Then here is the absolute minimum I need for the job: I need two devices, one with an on-off switch and two pushbuttons, and a second with two lightbulbs (see Figure 18). The buttons and bulbs are labeled "buy" or "sell." I retain the first device and give the second to my broker. Activating my device and pushing a button in Seattle causes one of the bulbs to flash on the one in New York, perhaps after a lag due to a finite signal velocity. The hope, then, is that there is a strong correlation—not necessarily perfect—between which button I push in Seattle and which bulb flashes in New York. Note that even with a less-than-perfect correlation I can still send the desired signal: I need only repeat the message many times.

Now suppose the device I keep in Seattle also has two labeled bulbs rather than a pair of pushbuttons. Furthermore, when I activate it, one of the bulbs lights, but not always the same one; in fact, the bulbs light

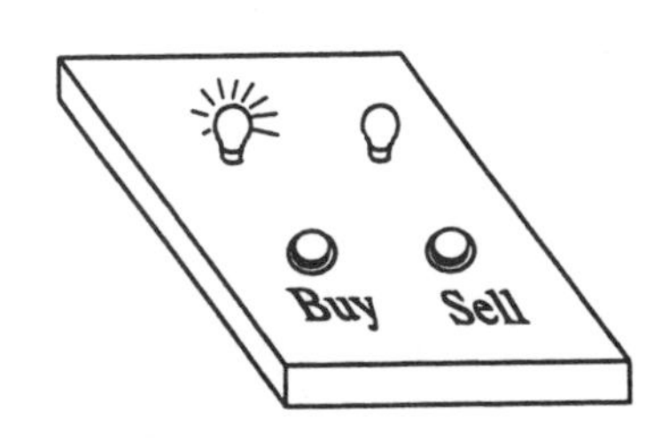

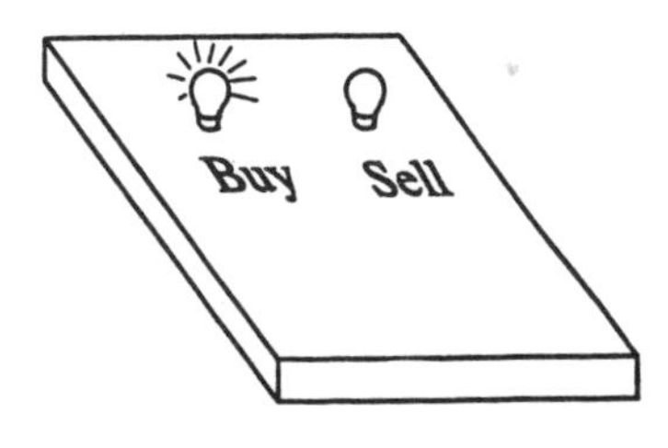

Figure 18

up seemingly at random. Then even if the correlation with New York is perfect, I cannot send a message—not without another means of communication to inform New York *when the bulb I wanted to glow actually lit up*. The perfect correlation is surprising but cannot logically be said to represent instantaneous signaling. This hypothetical situation is the one actually observed in the EPRB experiment; although an analyst for the Navy once recommended using the "EPR-effect" for long- distance communication with submarines, one cannot send signals faster than light with the EPRB apparatus. Bell, of course, understood this perfectly well in 1964. So what was the nature of the "nonlocality" he discovered?

Let us follow Bell's reasoning in his papers from the early 1970s. In Figure 19 two regions labeled "*A*" and "*B*" are shown on a space-time diagram. The shaded region is the intersection of the causal pasts of these two regions—their "joint causal past." We are interested in some measurement to be made in *A*; for instance, which way a particle was deflected by a magnetic field. Now suppose we are told the outcome of some measurement made in region *B* (e.g., "a particle was deflected up"). In this case it may be necessary to revise our probabilities for region *A*. If we have to do so, the revised probabilities are called "conditional probabilities."

Conditional probabilities are simple to understand. If I want the probability that my horse will win at Longacres, I total his past wins and

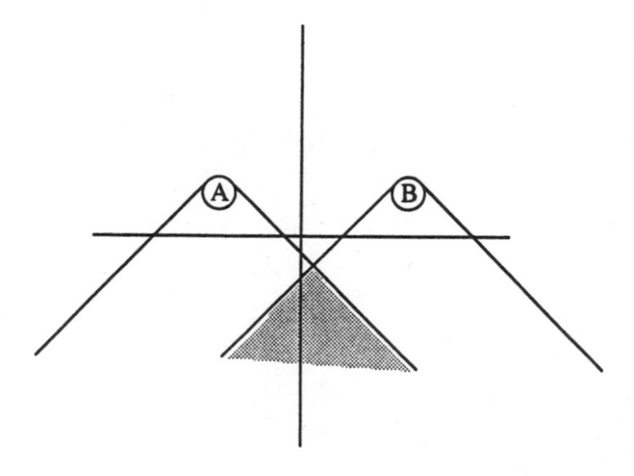

Figure 19

the races he has run at that track and divide. If I want the conditional probability that he wins on a wet track, I first throw out all the data from races run on sunny days and refigure. This recipe is just the obvious way to revise a probability in order to take account of some new information. If the conditional probability that something occurs, given that something else occurred, is the same as its usual (unconditional) probability, the two outcomes are called "statistically independent" or "uncorrelated." (The multiplication rule for probabilities holds in this case; see the Appendix.) Otherwise, they are "correlated." Statisticians caution us not to confuse the latter with either *logical implication* or *physical causation*, which may or may not be involved. Correlations can be spurious—I tend to doubt those claimed between sun spots and women's hem lines, or bird migrations and business cycles—but if real they at least suggest common causes.

Returning to Bell's discussion of locality, it is apparent that in the EPRB experiment the conditional probability that a particle is deflected up in region *A* (where the left analyzer is situated), given that the other particle was deflected down in region *B* (containing the right analyzer), is generally different from the unconditional probability. For instance, when the two analyzers are aligned, the conditional probability is one, while the unconditional probability is one-half. How could this correlation come about? Perhaps because the two regions have a common causal past. Information from this region may have propagated to both *A* and *B*—piggyback on the particles, say, or by some other means. But surely, reasoned Bell, if we took this common past into account, statistical independence would be restored. That is, *given a complete knowledge of events in their joint causal past*, any additional information about *B* should be irrelevant to events at *A*. This was Bell's new axiom, which I will call "local causality." (Bell was partly motivated in this work by experiments being planned; see Chapter 13. In real experiments, the statement "up on one side implies down on the other" will not hold precisely, the exact relation being replaced by a strong correlation; hence the need to discuss randomness.)

Bell's assumption of local causality may sound mysterious, but it is just a part of everyday reasoning about most situations. If a real-estate developer, an accountant, and the owner of a savings and loan all act in a way that ultimately transfers depositors' money into a Swiss bank account, we assume the explanation lies in a meeting between the principals. Each may sometimes act on whim, and chance may deflect their purposes—but we believe they must have followed a prearranged plan at least part of the time. Such an explanation we call a "conspiracy theory"; of course, in this era of telephones and fax machines, conspiracies can include (virtually) instantaneous signaling, but such communications are not required.

In 1971, Bell argued that realism plus local causality cannot explain atoms. Bell's argument this time was not completely rigorous and was criticized by Abner Shimony, Michael Horne, and John Clauser—the Americans referred to earlier, whom we shall meet in the next chapter. Clauser and Horne published the first rigorous theorem of this type in 1974; in 1976, Bell and the American group held a stimulating exchange by overseas post which appeared under the title "An Exchange on Local Beables" in a quantum-orthodoxy-doubting-subculture publication called *Epistemological Letters*. ("Beables" was a Bell neologism for existent things; it was criticized on etymological grounds since the suffix "able" refers not to existence but to potentiality, and so has not caught on even among members of QUODS.)

Once again there were hidden assumptions. Shimony, Horne, and Clauser jestingly suggested the following alternative explanation of EPRB. Two scientists decide to jointly carry out the experiment, with each to choose one of the analyzer orientations. However, unknown to them, a third party has prepared a list of analyzer orientations and experiment outcomes. This person sends the list of outcomes to the manufacturer of the particle detectors, who programs the apparatus in a way unknown to the experimenters to simply reproduce the outcomes on the list. The list of orientations is sent to the scientist's secretaries. Just as the experimenters are about to position the analyzers, the secretaries whisper in their ears, suggesting that they set the orientations to those listed. And voilà—the quantum-mechanical correlations are reproduced!

At issue, of course, is the free will of the experimenters. Might past events, and in particular those unknown quantities (the "hidden variables") that determine where the particles go, also fix the behavior of the experimentalists? Most of us are loath to consider this possibility: we ponder our future actions, which we are certain we direct, and we will never admit to being mere machines. If we replace the experimenters by devices—random number generators of some type, say—we might entertain this notion, but the scientists who discussed this issue in the 1970s agreed that this peculiar explanation should be ruled out by *fiat*.

On another issue, could not the real-estate developer, the accountant, and the banker mentioned above just happen to act in a way that enriches them all, giving the illusion of a conspiracy? The role of chance in Bell's theorem required additional clarification. In 1985, the Princeton mathematician Edward Nelson, the man who rediscovered the alternative theory based on Brownian motion, introduced some useful terminology in this area.

Nelson's contribution might have been prompted by the most quoted of all of Einstein's objections to quantum mechanics. Here is Einstein amplifying somewhat on his famous objection to a dice-playing God: "That the Lord should play with dice, all right; but that he should gamble according to definite rules, this is beyond me." (As Einstein worried more about the abandonment of *realism* than the overthrow of determinism—see Chapter 19—I admit this may have been just a riff on his standard joke.) By "gambling according to definite rules" Einstein meant quantum mechanics, but to a probabilist it sounds more like a description of a stochastic process. Nelson asked what assumptions might imply "locality" if the world is indeed such a regulated dice game. To answer this question, he first clarified several terms in the physicist's lexicon that have multiple meanings.

Consider first "observable," or "variable." This might mean anything we can observe in an experiment, but such variables come in two classes. The first consists of variables under our control. An example would be the orientation of the left analyzer in the EPRB experiment. Such controllable variables are often called "parameters." The second class consists of observables that fluctuate even if we exactly reproduce the experimental setup. The detectors in the EPRB experiment monitor such a quantity; it is customary to call these "random variables." By way of analogy, in an automobile trip the tire pressure you put in the tires is a parameter; how much time passes before one goes flat is a random variable.

Suppose that in our universe determinism fails: particles can spontaneously pop into existence, lightbulbs can spontaneously light up, and experimenters may even choose their experiments freely. What kind of "local causality" might still be imposed on this random world? To begin, Nelson asked what rule would eliminate signaling faster than light. To send a signal we must vary a controllable parameter; clearly, doing so should not affect events in the past of, or spacelike related to, our laboratory. So Nelson introduced the term "active locality"—AL, for short—for a situation in which varying a controllable parameter in a space-time region *A* can only produce an observable effect in the future light-cone (the "causal future") of *A*. This assumption prevents instantaneous signaling but leaves the question of residual correlations open. So Nelson introduced another assumption called "passive locality," or PL. Suppose we "fix" everything in some region of the past of our laboratory. (Admittedly, the term "fixing" is not completely appropriate here, since we have just agreed that some variables cannot be "fixed"; what is really meant is replacing ordinary probabilities by conditional probabilities, conditional on knowing everything, controllable or uncontrollable, in a suitable part of the past.) Suppose then that statistical independence between spatially separated laboratories is restored. (See Figure 20.) This axiom includes Bell's "local causality" as a special case, since we can take the past region to be the joint causal past of the two laboratories. But we might have taken the whole universe one nanosecond after the Big Bang.

In making these distinctions, Nelson zeroed in on a point I called to the reader's attention earlier. There may be "causes" in the universe that cannot be generated at will by the experimentalist. Consider again the two devices I could use to send "buy" or "sell" to my stockbroker in New York. Suppose both devices have two bulbs that light up randomly,

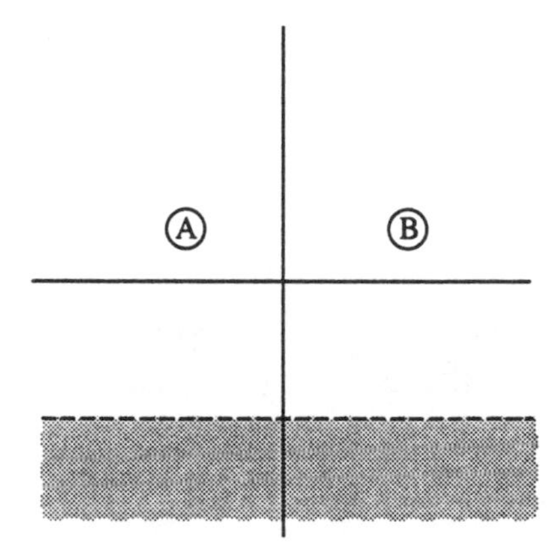

Figure 20

simultaneously, and show a perfect correlation. I might claim that my device is a "transmitter" signaling the stockbroker's "receiver" in New York; but my stockbroker would be equally justified in claiming that I have the causal order inverted, that the "cause" is in New York and the "effect" in Seattle. Without a time-order to distinguish cause from effect, neither of us can win the argument—but even given a definite time-order, causation need not be involved. Consider an extreme case: the two devices are temporally rather than spatially separated, perhaps being "the same device" lying on my desk on the first day of each month in 1990 and again on these dates in 2000. If the perfect correlation persists—the same bulb lighting on the same day of the month, 10 years later—we could call the later events the "causes" and the earlier ones the "effects" without invoking the Grandfather Paradox (but we would refrain from such language and simply call such a phenomenon *periodic*). Thus our understanding of cause and effect is intimately tied to our ability to control variables.

In 1985, Nelson proved a version of Bell's theorem which took the following form: realism plus AL & PL implies a mathematical relation violated by quantum mechanics. His theorem was stronger than Bell's (that is, it covered more ground), since the hypotheses are weaker. Nelson proved an inequality as usual, but it is simpler to show his axioms taken together imply determinism, thus recurring to the previous case—

Bell's 1964 proof. (See the Appendix for this proof; it assumes perfect correlations between "up" on the left and "down" on the right and so is less general than proving an inequality, which always leaves some "room" left over.)

From the realist's point of view, two hypotheses are better than one—denying the second one may be less painful. Nelson's theorem suggests abandoning passive locality while retaining active locality and the realism principle. Passive locality is a rather complicated restriction to state, and its denial may be compatible with relativity. But one thing is certain: *determinism must go*. For a deterministic theory is passively local by definition: if fixing all the variables in the past determines the future completely, there are no fluctuations left to be correlated.

Thus dice-throwing gives some hope of reconciling relativity and realism. If Einstein had known this, he might not have begrudged the Old One the right to roll the bones.

13.

Testing Bell

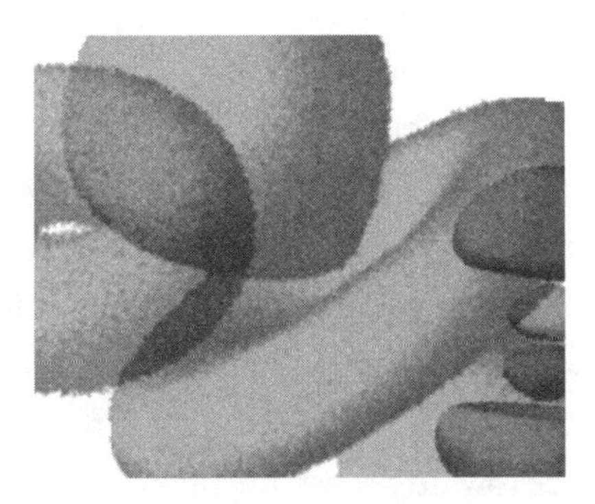

The [experiment] considered above has the advantage that it requires little imagination to envisage the measurements actually being made.

—John Bell, "On the Einstein-Podolsky-Rosen Paradox," 1964

The collaboration in designing the experiment was the high point of my intellectual life. It was also a great privilege to come to know John Bell, who was as admirable a man as I have ever met.

—Abner Shimony, 1993*

We had a great time.

—John Clauser, 1993†

Bell's assessment of the experimental prospects was optimistic: five years would pass before anyone thought of an experiment to test his inequality. Even then it would not be easy to get it to work.

Among the terms colleagues use to describe John Clauser are impatient, self-confident, and irreverent. Clauser was born in 1942 into a

*Personal communication to the author.
†Personal communication to the author.

virtual Caltech dynasty. Francis Clauser, John's father, and Francis's twin brother both acquired the B.A. and Ph.D. there, and other members of his family attended Caltech as well. The Clauser twins were students of von Karman, who had brought the modern science of fluid dynamics to Caltech in the 1930s. An early believer in the importance of studying difficult nonlinear effects, Francis Clauser had developed his own picture of atomic phenomena, based on fluid-dynamical descriptions inside and outside of particles. He was distraught when he could not understand the new and strangely linear theory called quantum mechanics. "We had 1,000 discussions starting when I was in high school," John recalled. The elder Clauser's attitude toward the establishment also impressed the future physicist. "Don't believe everything they tell you," Francis would caution his son. "Always go back to the experimental data—common wisdom is frequently a poor interpretation of what is actually observed."

"I grew up in QUODS (the quantum-orthodoxy-doubting-subculture)," Clauser says, "and went to Caltech asking all the questions people couldn't answer." The most popular physicist teaching at Caltech in that period was Richard Feynman, who occasionally carried his bongo drums over to fraternity parties. Clauser remarks that he must have sat in Feynman's lectures just before the famous "red books," the three-volume set of the *Feynman Lectures on Physics*, appeared, but he mostly recalled reading on his own. Besides, college life offered other diversions. Thus was a confrontation of the generations over the quantum theory averted. (Clauser used an unprintable epithet about Feynman's claims that quantum phenomena admit no classical explanations. About the two-slit experiment, the Californian remarked: "If a bunch of surfers pass through a breakwater with two entrances, you'll see the two-slit pattern later on the beach in surfer flesh!")

After college Clauser went to graduate school in experimental physics at Columbia. There he went to work for Patrick Thaddeus, a former student of Charles Townes (of laser fame), who groomed Clauser for a career in astrophysics. The hottest topic at the time was the newly discovered 3° Kelvin microwave background radiation, a fossil relic of the Big Bang. This became Clauser's thesis topic; his and Thaddeus's measurements were the third made, and the value accepted today still lies within their error bars. During this period, Clauser recalled, "I didn't think about quantum mechanics as a career. I just didn't understand it. I read de Broglie's book on nonlinear wave mechanics, a long review article by Hans Freistadt and Bohm's articles."

Then, sometime in 1967, Clauser spotted an odd-looking periodical on a shelf at the library of the nearby Goddard Institute for Space Studies. It was, he noted, an early issue of a journal with a peculiar title: the word "Physics" repeated in three languages. Idly riffling through the pages, he came upon Bell's article. "I realized it was important," Clauser recalled later, "but at first I didn't believe it. I went through a cycle of trying to find a counter-example, trying to prove it, trying to refute it. Then one day I realized there was nothing wrong. Jesus Christ, I thought, this is a very important paper. Why had no one else picked up on it?"

Bell's article was the primary stimulus for Clauser and the other experimentalists interested in realism, but it had an important precursor. In 1957, David Bohm and a student at the Technion in Haifa, Yakir Aharonov, published a paper entitled "Discussion of Experimental Proof for the Paradox of Einstein, Rosen, and Podolsky." W. H. Furry, a Harvard physicist, had discussed in 1936 a proposal of Schrödinger's that pairs of systems that become sufficiently separated might lose the quantum "entanglements" responsible for the EPR correlations. Aharonov and Bohm, after a concise discussion of EPR and Bohr's response—"For [Bohr] the system of two atoms plus the apparatus ... is, in any case, basically inseparable and unanalyzable, so that the question of how the correlations come about simply has no meaning"—described an experiment "verifying the paradox of [EPR]."

The experiment had been suggested by John Archibald Wheeler of Princeton in 1946 and was carried out in 1949 by Chien-Shiung Wu and Irving Shaknov at Columbia. (Wu is better known for her later and more famous experiment showing that elementary particles can possess "handedness," a discovery that arguably deserved a Nobel Prize.) Wheeler had proposed that the pair of photons generated by positron-electron annihilation could be used to test quantum electrodynamics. QED predicted that the two photons must have opposite polarizations much like the atomic spins in the EPRB experiment. (The property of light analogous to an electron's spin is called "polarization"; in the classical wave picture it means that the electric field vector has a definite orientation. Polaroid sunglasses reduce glare by passing only light polarized in one direction.) Wu and Shaknov caused the annihilation radiation to scatter off anthracene crystals, which served in place of polarizing filters for their very energetic photons. They measured the effect at 0° and 90° relative angle, and the quantum prediction held. This disproved the conjecture discussed by Furry, but unfortunately Aharonov and Bohm limited their analysis to this specific hypothesis. If Aharanov and Bohm had thought of finding a more general proof applicable to the whole class of local hidden-variable theories, they might have scooped Bell by a half-dozen years.

When Pat Thaddeus discovered that Clauser was not proceeding along the standard route for a bright young astrophysicist, he gave his student a stern talking-to. Quantum mechanics had all been understood years before, he told Clauser; either read it in Bohr and Einstein or talk to theorists. So Clauser did both. About an experimentalist's first visit with a high-powered theorist, Clauser remarks, "It's a coming-of-manhood thing. They 'know more' than you do and let you know it. But here I thought I knew more physics than they did." He went to see Robert Serber, one of Oppenheimer's students, and tried to explain Bell to him. Without visible effort Serber wrote the quantum-mechanical prediction for EPRB on the blackboard, and Clauser said: "Right. That's what I want to test." Serber shook his head; it is all obvious, where is the experiment? Clauser stood his ground. After a while Serber, getting exhausted or exasperated, dismissed him, saying, "It might be worth pointing out in a letter to the editor, but no good experimentalist would try to measure it."

Undeterred by the thought of becoming a "bad experimentalist," Clauser searched for a way to test Bell. He had read Bohm and Aharonov, so he went to Madame Wu, told her about Bell's paper, and asked whether she had measured correlations at angles other than straight and perpendicular. She had not; nor did Bell's theorem particularly interest her. She sent

Clauser off to talk with one of her students, Len Kasday, who was redoing the experiment. Clauser convinced himself that photons from positronium could not prove a violation of Bell's inequality because the polarizations of high-energy photons were only weakly resolved. (Physicists, so fond of coining euphonious Greek-or Latin-based phrases, invented "positronium" for the ephemeral element formed by an electron and a positron dancing round each other in the instant before annihilation.) But as similar if less severe problems plague all correlation experiments, it was still worth doing. Kasday announced results in favor of quantum mechanics in 1970 (the details appeared in 1975).

Intent on finding a definitive experiment, Clauser took the train up to Cambridge, Mass., where Dan Kleppner and David Pritchard of MIT were conducting experiments on crossed beams of alkali metals. Clauser had read a paper by Peres and Singer suggesting that scattering in perpendicular beams might produce suitably entangled states. He gave the MIT group a seminar on Bell's paper, and at the end Pritchard turned to a newly arrived post-doc, Carl Kocher, and said: "Carl, wouldn't your experiment test this?" Kocher replied, "Of course, that's why we did it." As a graduate student working under Eugene Commins in Berkeley, Kocher had performed two-photon, low-energy polarization-correlation experiments in 1966, without knowing of Bell's paper. Commins and Kocher had originally conceived of their experiment as a lecture demonstration for a quantum mechanics course—the quantum analog, perhaps, of the pop-gun-shoots-the-monkey contraption used to illustrate Newton's laws—but when the measurements proved more difficult than expected, it became Kocher's thesis. Now Clauser had a possible candidate for a definitive experiment—but had Commins and Kocher already generated the data needed to test Bell?

On returning to New York, Clauser looked up the reference and discovered to his amazement that Commins and Kocher had also measured only at 0° and 90°. (There was probably some talking past each other that day in Cambridge. Kocher may not have immediately seen the difference between EPR correlations—which he and Commins had demonstrated—and Bell's theorem.) Clauser next wrote to Bell, Bohm, and de Broglie.

All three replied, and each agreed with Clauser's hopeful assertion that a repeat of Commins-Kocher at intermediate angles would be convincing. Bell, who did not make any experimental proposals himself in these years, was particularly enthusiastic. (Bell recalled that the letter from the American student in 1969 was the first serious reaction he got to his paper—after a lag time of five years.) Emboldened by this survey of the known realists, Clauser sent an abstract outlining his experimental proposal to the organizers of the Spring meeting of the American Physical Society. When the conference bulletin appeared, he received an unexpected and disquieting phone call.

The call was from a physicist at Boston University named Abner Shimony. Shimony told Clauser that he and a student, Michael Horne, had arrived at the identical conclusions. Furthermore, they were already working with another student at Harvard on doing the experiment. Shaken by the thought of missing out on what might be the experiment of the century in atomic physics, Clauser agreed to a meeting on neutral ground (Washington D. C., at the APS meeting), to discuss matters.

A professor once warned Abner Shimony not to become "one of those philosopher-physicists." Shimony was born in 1928 in Columbus, Ohio, and raised Orthodox Jewish by his intellectual father and stepmother in Memphis, where the family moved when he was four. (He has since lost any Southern accent.) Shimony started in physics as an undergraduate at Yale in the 1940s, but switched to mathematics and philosophy partly because the latter "offered a grand perspective." In 1950, he took an M.A. in philosophy at Chicago and then returned to Yale to write a dissertation on probability and inductive logic under John Myhill, completing his doctorate in philosophy in 1953. Then, like John Bell in Ireland, he read Max Born's *Natural Philosophy of Cause and Chance* (and an article on probability by E. Hopf that appeared in 1934), and his perspective changed. He told his wife that he had finally found his true interest and wanted to go back to graduate school—this time in theoretical physics.

After completing a hitch in the Army, Shimony went to Princeton. There he learned advanced quantum mechanics from John Wheeler and relativistic generalizations of quantum mechanics from Arthur Wightman.

(Wightman, an American mathematical physicist, introduced axioms for relativistic quantum field theory in the 1950s; see Chapter 19, "Opinions.") Wightman's first assignment to his new student was to read the EPR paper and find the mistake. "I did not find the flaw, and to this day I believe it is a correct argument, given its premises," Shimony said 30 years later. Discouraged by the difficulty of the mathematics in quantum field theory (probably the most severe test of mathematical skill that exists in physics), Shimony turned to Eugene Wigner for a problem in statistical mechanics. Wigner recommended one, but as Wigner was studying the Measurement Problem at the time—unlike most physicists, Wigner did not think the Copenhagen interpretation had settled the matter—Shimony could not escape the lure of quantum philosophy. Encouraged by Wigner, he gave his first talk on the topic, entitled "On the Role of the Observer in Quantum Theory," at a conference organized by Boris Podolosky in 1962, with Bohm, Aharanov, Dirac, and Rosen in the audience. (Shimony used the word "solipsism" at one point in his talk. When Dirac asked what it meant, Shimony thought he was being set up—but it turned out Dirac had simply never encountered the term!) The same year Shimony was awarded a doctorate from Princeton in physics. He taught at MIT for a while and moved to Boston U. in 1968.

In late 1964 or early 1965, Shimony received a badly mimeographed copy of Bell's paper from a friend through the mail. Despite the unattractive appearance and some arithmetic mistakes in the manuscript, Shimony recalled thinking "that the author was on to something... and had a powerful intellect." Shimony began a search similar to Clauser's for an experiment testing Bell's theorem, finding Wu-Shaknov by way of Aharonov-Bohm, and Kocher-Commins through a tip from a friend at Princeton. But then the road became less clear. At a conference in New York, Shimony buttonholed Aharonov about Bell and received the unexpected response that Wu-Shaknov already adequately tested hidden variables. Shimony, worried that he had misunderstood something, set out to understand the details of scattering theory—causing a several-year delay in selecting an appropriate experiment.

One day while Shimony was pondering these matters, a colleague suggested he take on a student who had arrived in Boston in 1965 from the Deep South. Michael Horne was born in 1943 in Gulfport and majored in physics at the University of Mississippi. He came up north already fascinated by another foundational question in physics—the ensemble problem in statistical mechanics. (The analogue of the Measurement Problem in statistical mechanics is: why are the ensembles of Boltzmann and Gibbs those singled out by Nature? After more than a century, it remains unanswered.) In the fall of 1968, Horne walked into Shimony's office. It did not take Shimony long to size up the student. "I had been in Abner's office exactly five minutes when he showed me Bell's paper," Horne recalled. "He asked me to try and turn it into a real experiment."

Shimony assigned Horne the task of performing the quantum-mechanical calculations for the Commins-Kocher experiment. Coincidentally, Frank Pipkin at Harvard had set out with two students—Gil Nussbaum and Richard Holt—to use Commins-Kocher type apparatus to measure the lifetime of an excited state in mercury atoms. Shimony came across the river to talk about Bell, and Pipkin suggested to Holt that he set up an EPR experiment. Curiously, Holt had already heard of Bell; P. C. Martin, the instructor in his undergraduate quantum-mechanics class, had recommended the intriguing new article to his students soon after it appeared. Holt had the energy to locate the obscure journal in the library but was not sufficiently interested in Bell's paper to follow it up. (A decade later at the University of Washington, I heard about a physicist who wanted to trap a single elementary particle and displayed a similar inertia. Opportunity whispers and never insists.) When Pipkin later made his suggestion, Holt's reaction was that it would be nice to carry out such an experiment in a few months and then get on with the real subject of his thesis. Before setting up the experiment, Holt first made an important theoretical contribution to the problem of calculating the correlations for real collector apertures; Horne and Shimony had assumed for convenience that the detector apertures subtended an infinitesimal solid angle. This proved to be more than a mere technical issue (to be discussed in the next chapter).

By March 1969, Shimony and Horne had finished their theoretical preparations, but it was too late to report on them at the Spring meeting of the APS. Horne was apprehensive about waiting, but Shimony assured him that nobody else was working on "such far-out things." Then, like Darwin reading the letter from Wallace, they saw the bulletin of the Washington meeting containing Clauser's abstract. "We were stunned," Shimony recalled 30 years later. Uncertain what to do, Shimony turned to Wigner for advice. Wigner was sympathetic, pointing out that simultaneous discoveries were common—it had happened to him once—and suggested they call Clauser and propose joint publication. Clauser agreed at the APS meeting, and eventually they added Holt to the team as well.

In May of 1969 Clauser went up to Boston for discussions. One problem requiring attention was the perfect correlations assumed by Bell and EPR. No real experiment can prove such ideal relations, so the team—adding to the alphabet soup of this subject, this gang of four is abbreviated CHSH—came up with a new version of Bell's argument that did not require them. Now it was time to write their proposal. Clauser, who lived on a boat moored in the East River, had scheduled a post–Ph.D.-defense sailing trip down the coast that summer. As he proceeded, he stopped periodically to telephone Boston and argue points of phrasing; by the time he reached Florida, they had their preprint. It appeared in October 1969 in *Physical Review Letters*.

The Kocher-Commins experiment discussed by CHSH has the simple geometry shown in Figure 22. The source of the photons is a beam of calcium atoms produced by a tantalum oven at 1000 degrees. Under irradiation by intense ultraviolet light, the calcium atoms in the beam undergo a "cascade," a series of transitions in which an electron absorbs a photon and passes from one spherically symmetric state to another by way of an intermediate state, releasing a pair of photons in the process. Due to the symmetry of the initial and final atomic states, the two photons will have polarizations correlated analogously to the spins of the EPRB experiment. Since the photons are in the visible range, it was conceivable that ordinary Polaroid filters and photon counters (photomultiplier tubes) could serve as analyzers. Besides pointing out why previous experiments

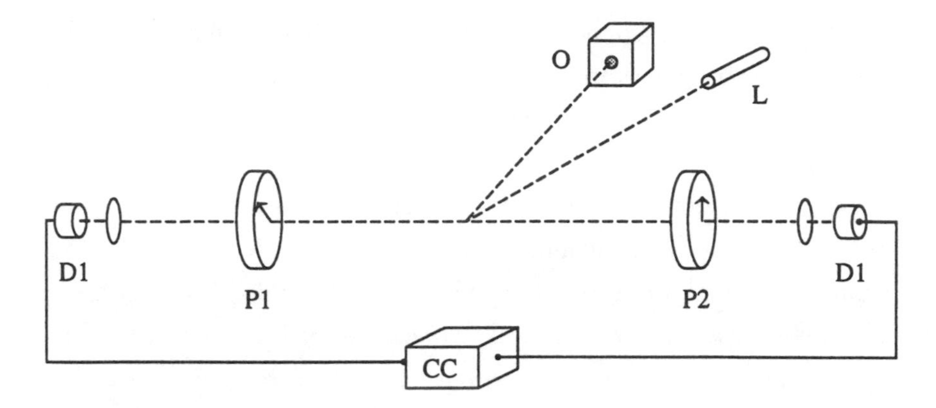

Figure 22

failed to test local realism, CHSH discussed how efficient the polarizers and detectors would have to be (very efficient), and of course remarked that data were needed at more angles.

Clauser next went to Berkeley on a post-doc to work on radio astronomy with Charles Townes. Since Commins was also at Berkeley, Clauser thought he might be allowed to spend some time on the proposed test of quantum mechanics. "This was not 'PC' [politically correct] at the time," Clauser recalls. Most physicists thought testing quantum mechanics to be an absurdity—sort of like "testing" the rules of calculus. So Clauser was pleasantly surprised when Townes agreed to let him spend half-time on the project, and Commins offered a graduate student, Stuart Freedman, to help with the experiment. Freedman, born in 1944 and educated in physics at Berkeley, does not recall having any particular interest in quantum philosophy. He simply thought the experiment might be interesting, although possibly tricky. "I was just a dumb graduate student at the time, and John was very enthusiastic," Freedman said later. "I had early trouble with him because he didn't realize it might be difficult." Clauser later admitted that Freedmen's complaint was justified. "I was impatient," Clauser said. "My spies in Boston [Horne and Shimony] told me Pipkin and Holt already had results."

At Harvard, Pipkin and Holt were agonizing over their experiment, allowing Clauser and Freedman to reach the finish line first; they published in 1972. Clauser, who calls himself a "hardened arch-critic" of the old school, genuinely hoped the experiment would prove quantum mechanics wrong. He accepted a bet from Yakir Aharonov at two-to-one odds against the quantum-mechanical prediction and told Freedman he would quit physics if it proved correct. The others in the game were less sanguine about overthrowing physicists' most accurate theory—but with the stakes so high, one could always dream.

The result of this first round of experiments was ironical. To Clauser's astonishment, the quantum formula worked perfectly (Figure 23) in his experiment with Freedman. (He mailed a two-dollar bill to Aharanov, who still has it on the wall of his office.) Holt's and Pipkin's experiment, although close to the boundary between quantum mechanics and local realism, nevertheless upheld Bell's inequality. They delayed announcing a result because they simply did not believe it. One explanation of the disparate outcomes of these experiments may have been the different commitment of resources: Clauser and Freedman had massive glass polarizer assemblies spaced 15 feet apart, with winches and pulleys to rotate them, while Pipkin's and Holt's experiment fit on a laboratory table. (Freedman remarks that his and Clauser's apparatus may have been "the last heroic optical device" built before the ready availability of lasers revolutionized

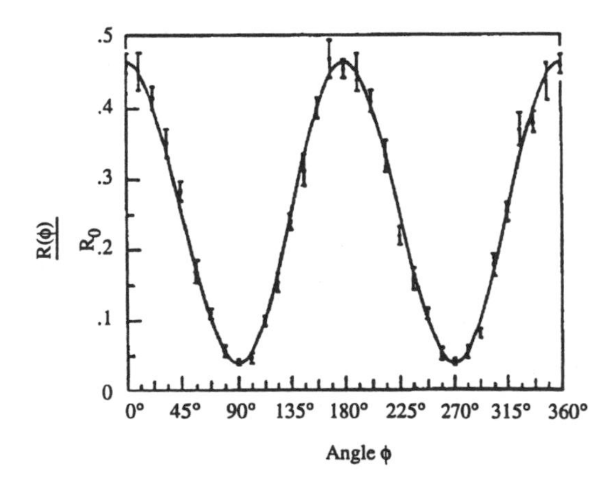

Figure 23 Reprinted with permission of Stuart Jay Freedman, Department of Physics, University of California at Berkeley, Thesis, 1972, p. 83.

the study of light.) But Pipkin and Holt also used a different photon cascade (in mercury rather than calcium), calcite rather than pile-of-plates polarizers, and an electron gun rather than UV light to excite the atoms.

Initially Pipkin and Holt refrained from publishing, although they circulated a preprint. (Some realists took brief comfort in their result.) Searching for the source of error in Holt's experiment—which Pipkin and Holt could never trace down—Clauser repeated it in 1976, with results favoring quantum mechanics. In the same year, Edward Fry and Richard Thompson at Texas A&M University performed another experiment using a different atomic cascade that also corroborated Clauser's and Freedman's results. In 1978, Shimony and Clauser—and, in the same year, Pipkin—reviewed the developments, bringing temporary closure to the campaign. Their conclusions are presented in the next chapter.

Reaction to news of the experiments in the physics community was mixed. Wigner remarked to Shimony that if these experiments had verified Bell's inequality, "it would have been very good for you, but bad for physics." Freedman recalled Louis Alvarez, a Nobel Prize winner, coming into his laboratory to make helpful suggestions, and other mainstream scientists such as Charles Townes and Eugene Commins expressed interest. Less sympathetic physicists, however, made snide remarks about Freedman and Clauser having confirmed the obvious. *Physics Today*, the trade journal of the American physics community, featured the experiment in an article and bolstered the participants' spirits somewhat by interviewing mere grad students for the piece. But everyone accepted philosophically that an experiment with heretical motivations that fails to contradict the reigning theory, however brilliantly conceived or carried out, does not get one on the boat to Stockholm.

A few comments about scientific life in America are in order. First, consider the question of collaboration versus competition. Sociologists of science debate "research styles" and the influence of extrascientific factors such as culture, race, and gender on scientists' choices. Sources for alleged "unhealthy competition" have been sought. Although a single instance proves nothing, and the scientists mentioned in this chapter—because of their interest in philosophy, their intelligence, and their skepticism—are certainly not representative of any group, it is interesting that when faced with a possibly unfriendly competition these four (white male) scientists preparing to test Bell simply joined together. Primitive competitive urges might be invoked to explain the later "race" between Clauser-Freedman and Pipkin-Holt to complete the experiment, but a continent separated them, and neither post-docs nor grad students usually have travel budgets.

Second, although the scientific establishment is by nature conservative, there is a tendency to exaggerate its power to create orthodoxy and suppress dissent. None of the scientists who performed the research described in this chapter expressed regrets over their involvement in controversial or "philosophical" issues. When asked if their participation in testing quantum mechanics adversely affected their standing among their peers, their ability to obtain research funding, or their job prospects, each replied simply "no." At least in the period of the middle 1970s, membership in QUODS did not imply ostracism from the physics community.

Where are these scientists located on the Bohr-Einstein axis? Except for Clauser, they profess middle-of-the-road views, neither pure naïve realism nor strict Bohrian positivism. A general feeling of unease with quantum mechanics was expressed, although Freedman suggested it is hubris on the part of physicists to imagine that they will always "feel comfortable" with everything they learn. Shimony has tried to see how much of realism can be maintained in a world containing strange entanglements. Holt thinks hidden variables may have been ruled out—but nevertheless quantum mechanics by itself cannot explain measurements. Horne described quantum mechanics as "begging for an explanation."

Clauser, a self-confessed "confused realist," stands out as the extreme skeptic of the group. Clauser is appalled that physicists accept a theory requiring an interpretation, like a piece of legislation. "Classical mechanics didn't need an 'interpretation'," he remarks. "There must be a paradox built in that people are trying to skirt." Like the late John Bell, Clauser believes the hard-liners falsely claimed to understand quantum mechanics. "If you asked 'why?' they said look it up in Bohr, and if you persisted they changed the subject or left the room," he remarks. "They didn't understand it either but wouldn't admit it." On the other hand, "Anyone who has thought about quantum mechanics is very disturbed. The first thing that happens is it makes you mad."

On one matter all the participants in testing Bell agreed: they hoped to see many-body quantum mechanics experiments flourish in the future. Each expressed a desire to know—in the words of a famous American theoretical physicist (see Chapter 16, "Paradoxes")—"how it could be like that."

14.

Loopholes

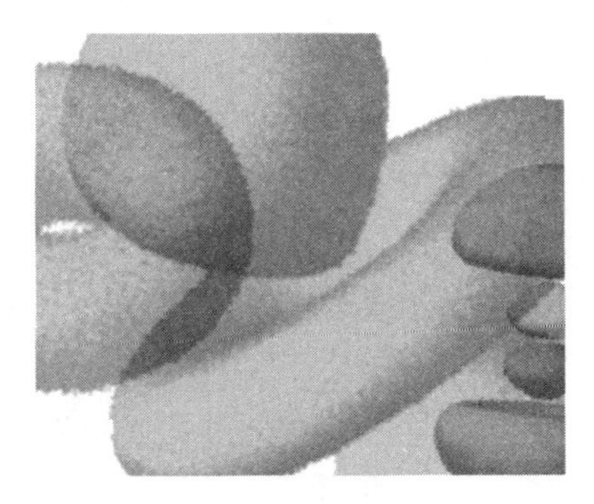

In 1985, David Mermin, a solid-state physicist at Cornell, wrote a funny and informative account of the EPR argument and Bell's theorem for *Physics Today*, the *Wall Street Journal* of the American physics community. Mermin's article generated a resurgence of interest in the subject, which swept like a cresting wave through the cocktail-party circuit and just as quickly receded. I recall confused discussions from this period with members of my generation, who alternately declared Bell's theorem trivial, uninteresting, or the greatest discovery since Einstein proved moving clocks go slowly. One sentiment produced universal agreement, however: do not write papers or apply for grants in the area, since "philosophical" issues are career killers.

Mermin made a pedagogical advance in the presentation of EPR by introducing "black boxes" equipped with switches and flashing lights—rather like the slot machines from "Bell's theorem." (I hereby acknowledge his priority.) After presenting a version of Bell's theorem and describing a recent variation on the Clauser-Freedman experiment performed in France, Mermin challenged the reader to provide *any* rational explanation of the EPRB correlations not invoking instantaneous signaling, other than that given by quantum mechanics.

Four months later, several readers did just that. In the Letters column, two groups of individuals—surprisingly, not including any of the Americans who carried out the analyses and experiments described in the last chapter—described perfectly correct, realistic models in agreement with the quantum mechanical predictions *and entirely free from action at a distance*. Although I cannot document it, I too discovered this strange explanation. Ignoring the advice of my cohorts, and having missed the reader's riposte in the letters column, I resolved to make my debut in the "EPR business" with an article for an appropriate journal.

My explanation was based on a flaw in the experimental arrangement, and so (being a mere amateur while Mermin was clearly a pro) I decided to run it past someone in the physics department first. It was not easy finding a physicist who would listen to a mathematician chatter about Bell's inequality, but soon enough I was writing out the details on a colleague's blackboard. As I was about to reach the punchline: "And so, you see, there is no need to invoke action at a distance!" the phone rang. My interlocutor picked it up, listened for a moment, exclaimed "My wife is having a baby!" and sprinted down the hall—leaving the lights on, the door unlocked, and all thoughts of quantum mechanics behind.

I left town a month later without revisiting my colleague, who was thus spared contemplating Bell's dilemma while changing a baby's diaper. I found another EPR enthusiast at the next institution, but before I had computed a single correlation he said, "Have you by any chance seen the letters column of last month's *Physics Today*?" And so ended "Wick's loophole," the clever explanation of the EPR experiment that everyone else had overlooked.

But the joke is on everybody, since the thing had been discovered in 1974 by Clauser and Horne and published in the not-particularly-obscure journal *The Physical Review*. ("Phys Rev," as it is fondly known, is the premier physics journal in the United States.) This is a depressing feature of modern science: the literature is so vast that virtually anything you can think of has already been published, but you have just not heard about it. Doing a literature search is always advisable, but considering that there are over 70,000 scientific journals—*titles*, not volumes—in circulation, this is often a hopeless task. (In the future, electronic publishing and computer database searching will improve this situation; in the meantime, publishing a scientific journal in paper form should be punishable by 20 years standing in line for the Xerox machine.)

Clauser and Horne based their realistic local model on an easily overlooked difference between the *Gedanken* experiment and the real experiment. Bell implicitly assumed that perfect or at least highly efficient particle detectors would soon become available. Unfortunately, this is very far from being the case even today; photon counters rarely have an efficiency better than 10 percent—that is, they miss 90 percent of the photons. (Polarizing filters are also not perfectly efficient; they sometimes reflect or absorb a photon even of the proper polarization.) If the detectors count or miss a photon independently of each other, only one photon pair in a hundred actually triggers both counters simultaneously. All versions of the EPRB experiment to date have allowed for this unfortunate circumstance by including a "coincidence detector": electronic circuitry that automatically rejects a signal from one analyzer unless the other one also registers within a certain "window in time," essentially the time necessary for the photon counter to reset after a detection. See Figure 22 in Chapter 13, "Testing Bell."

Hence we have "the loophole," as it is informally known (although there are others to be discussed below). Put in the form of an allegory, here is how it works:

Imagine that you are a spymaster "running" two agents, Boris and Natasha, who work independently of each other. You wish to send, once a day, a message instructing each agent to follow one of two plans, say plan "A" or plan "B." In order to cause maximum confusion for the FBI, you decide to arrange matters so that the messages received by the two agents will change randomly from day to day. On the other hand, you wish the agents to work in concert part of the time. Is it possible to find some scheme that will permit random results but still allow you to arrange *any* degree of correlation?

The answer is yes. Here is one procedure. You instruct each agent to randomly select one of three numbers each morning, say I, II, or

III, perhaps by drawing a circle divided into three regions on a sheet of paper and spinning a needle. You inform the agents that they will receive, once an hour starting at 5:00 A.M., a coded message on their Fax machine containing as a prefix one of these three numbers. They are to immediately reply "message accepted" or "message rejected" depending on whether the prefix agrees with the number they selected for that day. If they reject the message, they are to burn it immediately. If they accept the message, they are to wait for a confirming message: "decode"; if this is not received in five minutes, they are again to burn the coded document.

Now it is not hard to see that you can schedule for your agents to receive one of the two plans with any probabilities you like. For instance, you might wish to arrange the following: on a day in which Boris chooses I and Natasha chooses II, the probabilities are to be

Both receive A: 1/8. Boris receives A, Natasha B: 3/8.

Boris receives B, Natasha A: 3/8. Both receive B: 1/8.

You can arrange for these probabilities to be realized as follows. First, label a shoebox with the numbers "I, II" and put a large number of pairs of index cards (perhaps paper-clipped together) containing the encoded messages into the box with the indicated frequencies. That is, the pair of messages "A, A" should make up one-eighth of the total pairs in the box, "A, B" should make up three-eighths, and so forth. Make up nine boxes for the nine choices Boris and Natasha could make, each having the frequencies you desire. Now at the break of dawn, pick a box, select a pair of cards at random, and fax the messages on the cards to the corresponding agents, making sure to include the numeric prefixes. Wait for their replies. *If and only if* you receive two confirming messages, instruct your agents to decode the messages. Otherwise, pick another box and continue as above.

It is easy to see that with this procedure the agents will receive the plans with the desired frequencies. (If the spymaster wishes to enforce the choice associated with a particular pair of prefixes, he need only instruct the two agents to spin the dial again if the "decode" message is not received and then send messages chosen from one shoebox until both confirm.) The only drawback is that the spymaster receives two confirming messages only one-ninth of the time, so most of the messages are wasted.

I call this explanation the Spymaster Model, and I freely admit that it is a conspiracy theory (as defined in Chapter 12, "Dice Games and Conspiracies"). The correspondence between spy story and physics experiment is this: "plan A" or "plan B" corresponds to photon polarization states, and the three prefixes to three orientations of the polarization filters. Note that there is no instantaneous signaling in this scenario; the lapse in time necessary to transmit messages between the spymaster and his agents is built into the procedure. It is also built into the Clauser-Freedman experiment: the wire connecting the analyzers to the coincidence detector might carry analogous messages, although only the "photon detected/not detected" message is essential for the explanation to work. (Another perspicacious reader of Mermin's article wrote in to say that he could not understand Mermin's claim that "there are no connections between the pieces" of the apparatus, since there was a wire in the diagram of the experiment clearly connecting them!)

For the Spymaster Model to apply to the Clauser-Freedman experiment, either the glass polarizing plates or the photon detectors must discriminate against photons that carry the wrong "message." No one knows why glass plates might absorb photons of the "wrong type," or why photon counters might fail to count them. Nor do I have a new theory of light in which photons carry one of three messages. But these considerations are beside the point. Provided that at least eight of nine events registered by the counters are single detections, this crazy explanation cannot logically be ruled out.

So there is a loophole—but that is life in experimental science. No experiment is entirely "clean," embodying a perfect test of theory. The best version of Clauser-Freedman thus far (August 1993) was performed in the 1980s by Alain Aspect and collaborators at the Institut d'Optique Théorique et Appliquée in Orsay, a short Metro ride south of Paris. Aspect incorporated several improvements in his experiments. First, he had two-channel detection in each analyzer, capable of detecting the photon in either polarization state (previous experiments simply accepted or rejected the photon depending on polarization). Second, he invented a "switching system" that changes polarizer directions in less time than a light signal would need to propagate between the two analyzers, removing the possibility that an undetected signal passed between them (the so-called communication loophole). Aspect accomplished the latter by a neat trick: he used standing waves in water baths, which deflect the photons toward one or the other of the two polarizers in an almost random way. Aspect's data corroborated the quantum-mechanical formula.

Will improved photon polarizers and detectors soon eliminate the first loophole? Perhaps, but unfortunately there is more to the story. In March of 1991 Emilio Santos of the University of Cantabria, Spain, claimed in *Physical Review Letters* that photons from atomic cascades can *never* prove a violation of Bell's inequality, no matter how efficient the detectors and polarizers become. Santos backed up his claim by constructing a different local hidden-variable explanation of the Aspect-Clauser-Freedman experiment.

Several observations about photons from atomic transitions will serve to make this third (and even worse) loophole seem less implausible. (No parable or analogy is available for this one, unfortunately; hard-core quantum calculations are involved.) First, the source of the photons in the Aspect-Clauser-Freedman experiment is a three-body system, unlike the two-body source assumed in EPRB, since in addition to the two photons there is the source atom. Second, although the photons do have polarizations with suitable correlations *if no constraints on them are imposed*, there is an obvious one: to be recorded, both must enter the detectors. In effect, the restriction imposed on the photons—because of the spherical symmetry of the atomic cascade, most go flying off in "wrong" directions—culls the ensemble as in loophole 1 and changes the correlations.

To quote a noted baseball pundit, it was *déjà vu* all over again: the Americans had published this explanation, too. In 1974, Clauser and Horne noted these problems and found a different argument that circumvented them. They assumed that putting a polarizer before a photon counter never increases the probability of detecting a photon, no matter what "message" it might carry. Unfortunately, this property cannot be directly verified, because there is no known way to check when an atom has just emitted a photon. (All known methods of detecting photons absorb and hence destroy them.) With this "no-enhancement" hypothesis, Clauser and Horne proved a new version of Bell's inequality that ruled out both loopholes—but the extra assumption is not a consequence of realism or locality.

Strangely, Santos was one of the readers who straightened out Mermin in 1985—and then, five years later, committed a similar lapse himself. (Santos published an erratum a few months later crediting Clauser and Horne. I told many colleagues in the 1980s that Aspect's experiment proved that local realism is dead—so shame on me, too.) Santos did perform a service by reminding the community of the true experimental situation, however.

Now that the rediscovery of the work done in the 1970s is (hopefully) completed, one might ask whether this pattern is inevitable in science. Surely scientists should read more of the history of their subject before rushing into print? What about referees and editors at journals—aren't they supposed to prevent this kind of thing? Granted that omnipotence is impossible, one might expect a reviewer to check a few references in every paper, even if it means a stroll down to the library. Such care would help prevent redundancy in the literature, eliminate false self-assignment (usually unwitting) by authors, and maintain the historical continuity of a discipline. "Lately we seem to have been forgotten," Michael Horne remarked wistfully to the author; better refereeing might have obviated his complaint.

The experimental tests of Bell's inequality suffer from the same disease as the theoretical "no-go" theorems: hidden assumptions. To interpret real experiments, many assumptions are necessary, and listing all of them is as difficult as it is for theories. (Recalling Einstein's remark that theories determine what experimentalists measure, the problems could scarcely be less.) A cycle of proposing experiments, performing them, and then discovering loopholes is perhaps unavoidable. When the little community of Bell aficionados in the United States realized that photon correlation experiments could neither overthrow local realism nor falsify quantum mechanics, they wrote reviews and went on to other things. ("We're not just Bell physicists," Horne remarked to the author. Indeed, the experiments performed by these scientists were interesting tests of many-body quantum mechanics even without reference to philosophical concerns.) In the 1980s in Europe and the United States, a new generation went through another loop in the cycle—but we are all only human.

In the 1990s, new proposals for experiments surfaced, and there is some optimism that a definitive experiment is possible. In 1989, Daniel Greenberger of City University in New York City, Horne, and Anton Zeilinger of the Österreichischen Universitäten in Vienna proposed a version of EPR using *three* particles. Their version of Bell's theorem had the surprising property of not depending on statistics—if certain events are always realized, local realism is ruled out. In a later paper with Shimony, the authors argued that experiments using three photons may test their conclusions. Other experiments have been proposed that replace atomic spin or photon polarization with information about which arm of a two-arm interferometer a photon traversed. (An interferometer is a diamond-shaped arrangement of beam splitters and mirrors that creates two or more paths for a photon to take.) Interferometers have even been built that use *whole, neutral atoms* instead of photons. Edward Fry at Texas A&M has proposed carrying out a version of the original EPRB experiment; it may have the best chance of finally proving a violation of Bell's inequalities.

As accountants do for new tax bills, when these experiments are completed physicists will play "find the loophole" once again. "I have always had a queazy feeling that those assumptions will *never* be [verified]," Michael Horne remarked, "although Abner [Shimony] thinks otherwise." But, Horne continues, "by tinkering with multiparticle experiments maybe we can learn why quantum mechanics is here. As Rabi [the experimental physicist and Nobel Prize winner] used to put it when someone would bring it up: "Who ordered this quantum mechanics?'

15.

The Impossible Observed

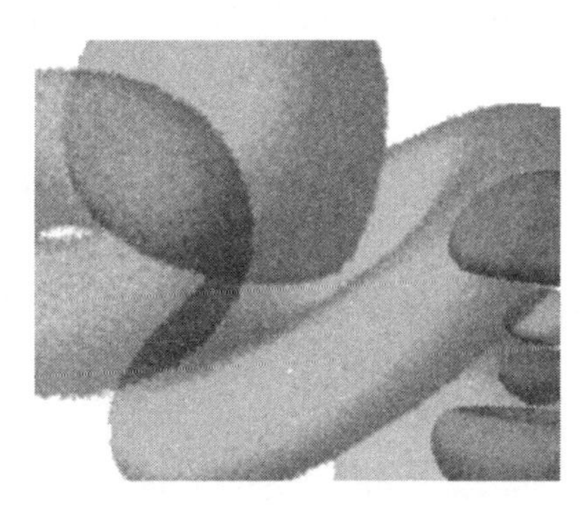

We never experiment with just one electron or atom... any more than we can raise Ichthyosauria in the zoo.

—Erwin Schrödinger, 1952

Here, right now, in a little cylindrical domain... in the center of our Penning trap resides positron (or anti-electron) Priscilla, who has been giving spontaneous and command performances of her quantum jump ballets for the last three months.

—Announcement from Hans Dehmelt's laboratory in Seattle, 1984

In 1979 in Heidelberg, Germany, Dr. Werner Neuhauser looked up from the eyepiece of a low-power microscope and fancied he heard the ghost of Mach whispering: "Now I believe in atoms." Neuhauser had just glimpsed what appeared to be a bright blue star floating in the void; it was a single barium ion, caught in an electromagnetic trap and fluorescing in a laser's beam. Thus transpired the first observation of an isolated atom using a lens; soon one would be glimpsed with the naked eye as well. (When I was a child, elementary science textbooks claimed that no one would ever see an atom with the naked eye. The authors had erred by assuming that *smallness* was the important issue; actually, *brightness* and *isolation* from other atoms are what matters. The laser-stimulated barium atom produces 10^8 photons per second; your eyes can collect several thousand. The normal retina is sensitive to even a few photons, so you can see the atom, just as you would a distant star or any other bright, isolated object.)

Neuhauser collaborated on the 1979 experiment with M. Hohenstatt and P. E. Toschek of the University of Heidelberg and H. Dehmelt of the University of Washington in Seattle. Hans Dehmelt in particular had a lifelong dream of capturing and suspending in space a single atomic particle; he had been promoting the idea for more than a decade, against considerable skepticism from the physics community. His quest for the perfect atomic measurement was rewarded with the Nobel Prize in 1989. Dehmelt's story is also directly relevant to our concerns, as his work produced the only direct clash between Bohr's philosophy and experimental reality to this date.

Hans Dehmelt emigrated to the United States in 1952. Dehmelt's interest in confining an elementary particle had been piqued in his student days in Germany, when his teacher drew a dot on the blackboard and declared "Here is an electron...." It had been thought since Dirac invented his relativistic wave equation in the 1930s that the electron had no structure at all and was literally a mathematical point. Capturing and controlling this most infinitesimal entity would represent a *coup de maître* of experimental physics. With D. Wineland and P. Ekstrom, Dehmelt first bagged a solitary electron in 1973, trapping it in an invisible cage of electric and magnetic fields. But a few more years were to pass before he and his collaborators in Seattle learned how to cool down and interrogate their prisoner. Once that was accomplished, they could study it at their leisure.

Of special interest was the electron's gyromagnetic ratio or "g-factor," essentially the ratio of the electron's intrinsic magnetism to its intrinsic "spin." Dirac had predicted that this (dimensionless) quantity was exactly 2, but the Dehmelt team found experimentally

$$\frac{g}{2} = 1.001159652188(4),$$

with the "4" in parentheses indicating that the result is uncertain—this is old-fashioned experimental, not quantum, uncertainty—by plus or minus four in the last decimal place. In other words, Dehmelt's team measured this number to *four parts in a trillion*. Remarkably, theory made an equally precise prediction. The theorist T. Kinoshita, using the mathematical recipes provided by quantum electrodynamics (the theory of Feynman et al. from Chapter 10) together with high-speed computers, had predicted this number to be

$$\frac{g}{2} = 1.001159652133(29),$$

with the uncertainty also due to experimental error (in the measurement of another constant, the "fine-structure constant," needed for the calculation) and unavoidable computational errors. The difference between these two numbers is tiny but possibly still significant: on the basis of the discrepancy—a few parts in a hundred billion—Dehmelt conjectured that the electron has a nonzero radius of about 10^{-20} cm. Even rejecting this claim, Dehmelt's measurement yields an upper bound on the electron's size three orders of magnitude better than that attained by the high-energy physicists, with their immensely powerful (and extremely expensive) particle accelerators.

In 1984, the startling announcement quoted above was issued by the press office of the University of Washington in Seattle. Dehmelt, working with R. S. Van Dyck, Jr., and P. Schwinberg, had performed the even more astonishing trick of confining a *positron*, the antimatter twin of an electron, for a period of months. Their positron originated in the decay of a radioactive nucleus, and they could be sure of its identity during those many days since it had no opportunity to exchange places with another positron. As time went by and their caged particle continued in existence, its spin-flips registering on their devices like so many turns of the wheel in a hamster's cage, they began to think of it as a pet—and, as one does for pets, gave it an affectionate moniker ("Priscilla"). The ghosts of Schrödinger and Mach had been properly propitiated.

In 1989, the Nobel Committee recognized these accomplishments by awarding the Physics Prize jointly to Dehmelt and two other experimental physicists: Norman Ramsey of Harvard, whose work led to the development of atomic clocks accurate to one part in 10 trillion, and Wolfgang Paul of the University of Bonn in Germany, who built the first radiofrequency ion trap (used by Neuhauser and collaborators to confine the barium ion) and contributed to the development of the mass spectrometer.

Dehmelt's measurements refuted claims made originally in the 1930s on grounds of principle. Niels Bohr, his disciple Wolfgang Pauli, and other proponents of the Copenhagen interpretation thought they had conclusively demonstrated that the magnetic moment of a free electron—one not bound in an atom—could never be observed. Bohr's argument was widely accepted in the 1970s; some theorists were supporting it as late as 1985. As today this quantity may be the best-measured number in all of science, Bohr's "proof" must take its place in the entrance hall of the Impossibility Hall of Fame, ahead of such celebrated demonstrations as the impossibility of heavier-than-air flight, space travel, or (for that matter) seeing an atom with your unaided eyes.

Before discussing Bohr's blunder, I present more details of Dehmelt's fantastic achievements. The Penning trap—named after the Dutch physicist Frans Michel Penning, who showed in 1936 how to trap an electron cloud in a discharge tube—which Dehmelt used to confine a single electron is sketched in Figure 24. It incorporates two metal caps, each carrying a negative charge, and a metal ring carrying an equal amount of positive charge. (For Priscilla's cage, the polarity of the charges are reversed.) The cap-to-cap distance is slightly less than a centimeter. The electrodes create a weak electrostatic field, which partially confines the electron. There is also a strong magnetic field—about 80,000 times the strength of the Earth's field; more than sufficient, I was warned, to erase the data on bank cards in my wallet—produced by a superconducting electromagnet (not shown). The whole arrangement is placed in the highest vacuum attainable and cooled with liquid helium to four degrees above absolute zero.

An electron moving in the combined fields of the Penning trap, *if thought of classically*, has the roller-coaster trajectory shown in Figure 25. The electron's motion is a sum of three motions: a vertical oscillation

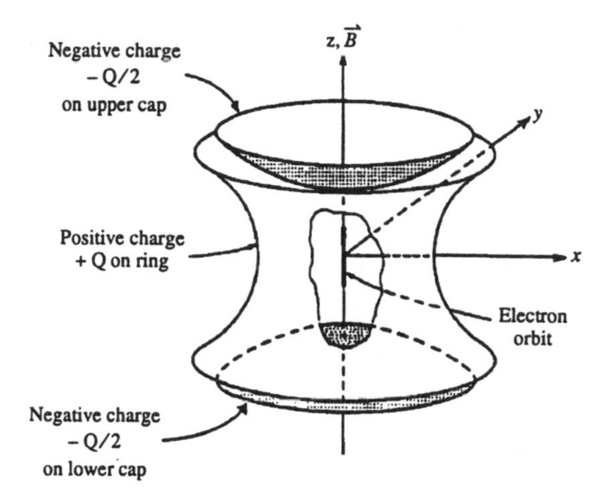

Figure 24 Reprinted from *Advances in Laser Spectroscopy*, E. T. Arrecchi, F. Strumia, and H. Walther, eds., pp. 153–87 with permission of Plenum Publishers, New York.

along the axis of the trap, limited by the electric fields (which act to form a "valley" with a "hill" at either end); a very fast "cyclotron" motion of the electron in small circles around the magnetic field lines; and an additional motion that slowly sweeps the tiny cyclotron loops around in a larger horizontal circle, completing the roller-coaster ride.

According to orthodox theory, electrons do not have trajectories. Whether true or not, it is certain that classical mechanics does not do justice to the electron's "real motion." The electron in the Penning trap forms a kind of nearly macroscopic "atom," which has been dubbed geonium. (Dehmelt explains that he adopted this term since the electron can be thought of as bound to the Earth by the combined electric and magnetic

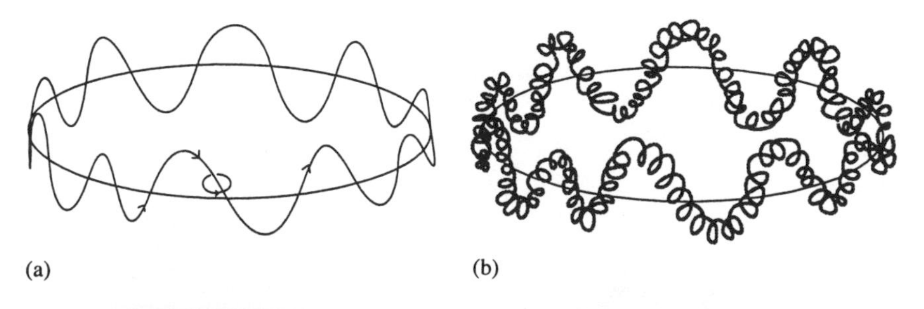

Figure 25 (a), (b)

fields of the Penning trap.) All three classical motions have corresponding quantum levels, and the electron performs Bohr's quantum jumps from one level to another. Dehmelt's team watches these on a voltmeter; they are seeing individual quantum jumps in an artificial atom. Geonium must be described not by Schrödinger's wave but by the more accurate Dirac version, which incorporates spin as well as relativity effects. The increase in mass predicted by relativity theory for a moving particle, for instance, is readily observed.

Beyond the imagination needed to design the trap, there was the hurdle to overcome of cooling a hot particle. This Dehmelt's team accomplished in stages. First, they cooled the vertical vibration by connecting a radio receiver tuned to the electron's oscillation frequency to the cap electrodes, siphoning off energy. The cyclotron motion decays spontaneously of its own accord to "ground" level by the emission of photons. Damping down the remaining motion, however, required special ingenuity. To accomplish this task Dehmelt used a method termed "radio-frequency side-band cooling," a neat trick most easily explained for a trapped ion.

An ion moving along a straight line is irradiated by an electromagnetic wave; see Figure 26(a). The wave's frequency is purposely set at a value slightly too low to allow the ion *at rest* to absorb a photon. The moving ion sees the frequency as slightly higher, due to the Doppler effect, Figure 26(b). (This is the effect that causes a train's whistle to apparently change pitch as the train passes.) Hence it can absorb a quantum of

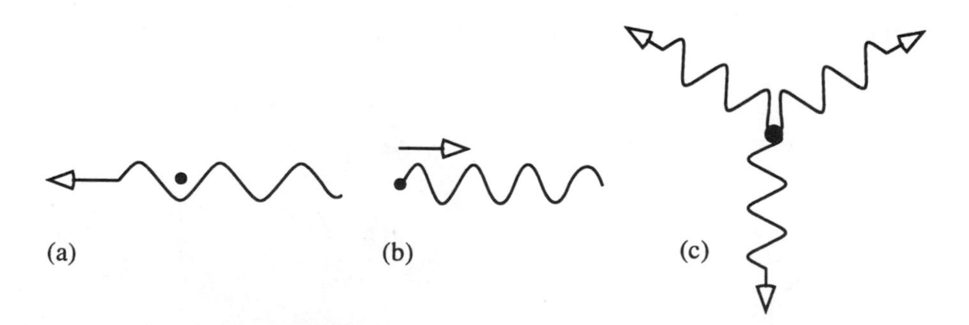

Figure 26 (a), (b), (c)

energy from the wave, which causes a momentum change or "kick" opposing its motion. When the photon is reradiated, it emerges in a random direction, Figure 26(c), causing no net change in the ion's momentum, statistically speaking. The ion, "cooled" by these techniques, settles down for a lengthy stay in its little cage.

To a mathematician, the ultrahigh vacuum needed to sustain the life of a trapped particle seemed the most remarkable technical achievement of Dehmelt's team. This is particularly striking for a trapped *positron*, since it will commit murder-suicide with any passing electron. But Dehmelt shrugs off the compliment; such vacuums, he remarks, are routine with modern-day cryogenic technology. At the very cold temperatures at which they maintain the trap, essentially everything but helium freezes out. The remaining molecules are adsorbed by a standard item called "thirsty glass" that serves as fly paper for molecules. Nature may abhor a vacuum, but to modern-day technologists, achieving a perfect one is apparently a triviality.

To observe the electron's spin, Dehmelt wrapped a nickel wire around his trap. The wire produces a small distortion in the magnetic field, causing the frequency of the vertical oscillation to become dependent on the electron's spin direction. Monitoring this frequency on an oscilloscope screen allows the detection of individual spin-flips; see Figure 27. In 1986, Neuhauser, Dehmelt, and colleagues observed seemingly instantaneous

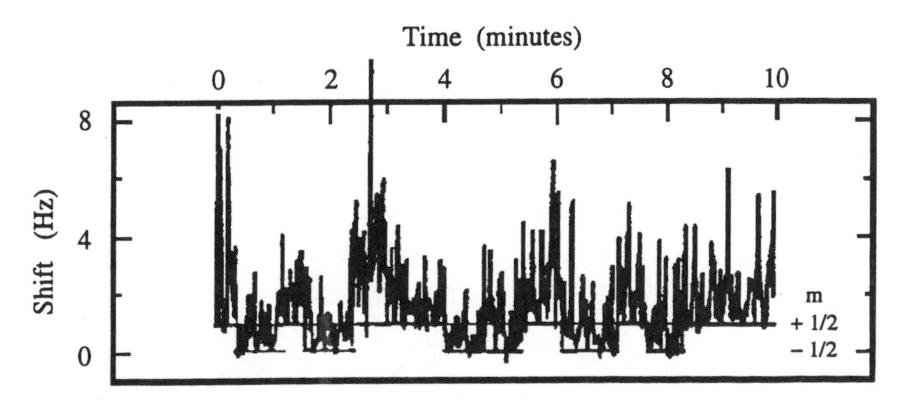

Figure 27 Reprinted with permission of *Physica Scripta* **T22**, p. 104.

transitions from "bright" (fluorescing) to "dark" states in a trapped barium ion. One wonders what Schrödinger ("those damned quantum jumps") would have said about these developments.

At this point I suspect the reader has become irritated by the role electron spin has played in our story, considering that no description of what it is supposed to be has been produced. And according to Dirac, the electron has zero radius—but how can a mathematical point exhibit "spin"? (Dehmelt's proposal of a nonzero radius for the electron was *not* intended to restore the classical picture of it as a little spinning marble.) Here is the modern physicist's understanding of spin: the point electron exhibits an intrinsic circular motion of less than atomic dimensions *at the speed of light*. This picture was established by Kerson Huang in 1952 using Dirac's theory. The apparent contradiction with relativity is resolved, or at least covered over, by the usual uncertainty principle arguments showing that the details of this motion cannot be observed—just as Heisenberg argued that the electron's orbit in the hydrogen atom is unobservable. Additionally, attempting a precise measurement of the electron's position would likely produce pairs of positrons and electrons—a characteristic feature of Dirac's theory—which would mask the original electron's position or annihilate it altogether. Adopting this picture of spin, the electron in Dehmelt's cage moves in a combination of *four* circles—shades of Ptolemy's epicycles!

Now I return to Bohr's claim that the spin of a free electron would never be observed. Bohr first formulated this notion around 1928, the year Dirac's theory appeared. The spinning electron hypothesis had been put forward in 1925 by R. Kronig to explain the splitting of certain spectral lines, but he let himself be talked out of it by Pauli at least partly because, if the spinning electron had the radius some people calculated for it, a point on its surface would move considerably faster than light—an ironical argument given the present picture. Consequently, the Dutch physicists Samuel Goudsmit and George Uhlenbeck, who received better advice from Ehrenfest, were able to publish the idea. Bohr's work was prompted by a paper of Leon Brillouin, who examined various possibilities for measuring the magnetism of a free electron, including shooting electrons straight at the north pole of a magnet. (The ones with spin pointing toward the magnet would be repelled.) He concluded that success was unlikely but not impossible. Bohr, however, had a different view. "It does not seem to have been generally recognized," wrote Bohr in 1929, "that the possibility of a direct observation of the magnetic moment of the electron would be inconsistent with the fundamental principles of quantum theory."

With the exception of Pauli, few physicists understood what Bohr was talking about. Part of the trouble was Bohr's lecture style: his assistant Rosenfeld described Bohr's talks as "masterpieces of allusive evocation of a subtle dialectic," which audiences found hard to follow. Here is Rosenfeld's description of a lecture given by Bohr in 1931 about spin:

> He had begun with a few general considerations calculated, no doubt, to convey to the audience that peculiar sensation of having the ground suddenly removed from under their feet, which is so effective for promoting receptiveness for complementary thinking. This preliminary result being readily achieved, he eagerly hastened to his main subject and stunned us all (except Pauli) with the nonobservability of the electron spin. I spent the afternoon with Heitler pondering on the scanty fragments of the hidden wisdom which we had been able to jot down in our notebooks.

The general drift of Bohr's reasoning was not difficult to follow—it was the details that baffled Rosenfeld and friend. Recall that Bohr based the complementarity principle on the assertion that all experiments must ultimately be described in classical terms. Electron spin, however, had no classical analogue. (Pauli originally opposed the use of the suggestive term "spin" for what he called "peculiar not classically describable two-valuedness" and thought of as something almost mystical.) True, for an electron bound in a magnetic atom, the electron's spin did make a contribution to the atom's overall magnetism, as Stern and Gerlach had shown. But the principles of quantum mechanics prevented this decidedly unclassical phenomenon from being detected in a free electron. "This," recalled Rosenfeld, was "the point [Bohr] ineffectually tried to make in his talk."

Luckily, one of the physicists in the audience, Nevill Mott, could make sense of the details and published the argument a year later, with Bohr's blessing. Mott and H. Massey incorporated the argument in their well-known text, *The Theory of Atomic Collisions*, whose third edition appeared in 1965. It is typical of arguments based on uncertainty, although a trifle more complicated than most: an electron is made to pass through an inhomogeneous magnetic field as in the EPRB experiment. Since the electron is charged, it experiences a force whenever it moves perpendicular to any component of the magnetic field. (The force on a charged particle in a magnetic field, first described by the Dutch physicist H. Lorentz in the last century, acts perpendicularly to both the particle's velocity and the direction of the magnetic field lines.) One tries to minimize this effect by shooting the particle straight down a field line, but the tilting of the lines—they cannot be perfectly aligned since the field is spatially varying—combined with uncertainty in the particle's position yields a force that perturbs the particle's trajectory. This perturbation is just sufficient to produce a blurring of the pattern on a photographic plate one hopes to observe. In other words, quantum uncertainty is mated to classical pictures as usual to yield a "no-go" theorem for a free electron in the conventional geometry of the Stern-Gerlach experiment. No doubt one really cannot carry out a spin measurement *in this unimaginative way* with free electrons.

Unfortunately, Mott and Massey jumped to an unwarranted conclusion: "From these arguments we must conclude that it is meaningless to assign to the free electron a magnetic moment." Mott and Massey, like von Neumann, Bohr, Pauli, and others in this century, fell into a trap set for those who desire to prove their ideology. One has first to imagine all theoretical or experimental possibilities contradicting the desired conclusion and then rule them out. Inevitably there are those of which the prover has not the slightest inkling. It may be unfair to ridicule these theorists for failing to foresee that one day an experimentalist would trap a single electron and force it to display its repertoire, as though it were a dancer at an audition, hoping for a part. Dehmelt did not need to detect the deflection of an electron in a magnetic field when a spin-flip occurred, since he measured the change in frequency of its orbit instead. (Dehmelt, who respects Bohr's views, believes he knows where Bohr erred. In his Penning trap, a wave-packet much smaller than the dimensions of the trap moves approximately along a classical trajectory *without dispersing*, thanks to the trapping fields. The nearly classical orbit is, he believes, the possibility Bohr and Pauli overlooked.) The moral is clear: be modest about the implications of your theories—and never underestimate the cleverness of the experimentalists.

Additional remarks. The experiments described here and in Chapter 13, "Testing Bell" were funded by government grants—that is, paid for by the taxpayers. The contrast between these brilliant table-top experiments and the equally brilliant ones carried out by the high-energy physicists can be stated simply in monetary terms: one particle accelerator costs 10,000 EPR apparatus or Penning traps. Perhaps we would get more for our dollars if we increased our support for those who still practice "Small Science."

16.

Paradoxes

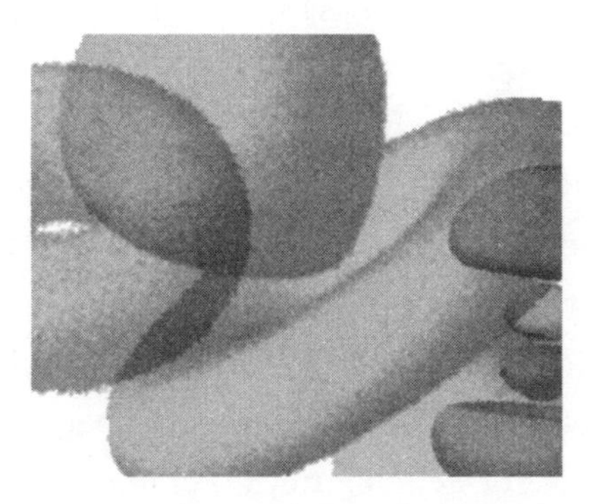

I think I can safely say that no one understands quantum mechanics... Do not keep saying to yourself, if you can possibly avoid it, 'But how can it be like that?' because you will get 'down the drain,' into a blind alley from which nobody has yet escaped. Nobody knows how it can be like that.

—Richard Feynman

As an earthquake has a main shock and secondary shocks, so too a scientific revolution. In this chapter I discuss some of the lesser but still significant tremblers that shook the foundations of physics since 1925 (the principal shock having been identified by Bell in 1964, following clues left by EPR and Bohm). Then, in Chapter 18, ignoring Mr. Feynman's mixed metaphor, we attempt to understand "how it can be like that."

"SPOOKY ACTION AT A DISTANCE"

This was Einstein's colorful phrase for the feature he found most objectionable in Max Born's interpretation of quantum mechanics. Imagine a single photon approaching a large piece of photographic film, say a square one meter on a side. Quantum mechanics might describe the state of this particle by a wave spread out over the whole square meter. But, as there is only one photon present, only one grain in the photo emulsion can be exposed, by a silver atom absorbing a quantum of light energy. After this event, the wave representing the photon's state must have collapsed to atomic dimensions. (To appreciate this change of scale, note that the area covered by one silver atom is to one square meter roughly as the dot at the end of this sentence is to the North American continent.) The photon has, so to speak, been sucked into the atom, traversing the 10 orders of magnitude separating macroscopic from microscopic in no time at all.

If the wave merely summarized our ignorance of the photon's true location—probability, in a classical interpretation, just serves to quantify such ignorance—there would be no problem. The "collapse" would simply represent an updating of our knowledge. But the Copenhagen view is that the wave represents the *complete physical description* of the photon's state. If so, the collapse instantaneously changes that state a meter away from the exposed grain—hence "spooky action at a distance."

It is true that one cannot use the collapse for instantaneous signaling (similar to the EPRB situation), because the wave itself is intangible. But how can something so ghostly represent the real state of affairs?

SCHRÖDINGER'S CAT

In 1935, Schrödinger, prompted by the EPR paper, wrote a lengthy essay—in a footnote he called it his "general confession"—setting forth his views on quantum mechanics. In a style that can only be called sardonic, Schrödinger criticized the strange opinions that hovered around his wave equation, like flies over a piece of meat. He began by summarizing the position most physicists held before quantum mechanics about what theory is supposed to accomplish.

"In the second half of the previous century," Schrödinger wrote, "there arose... an ideal of exact description that stands out as a reward of a centuries-long search and the fulfillment of a millennia-long hope, and that is called classical." This is the physics of *models*. Physicists tried to grasp reality by use of an *image* or *model*, says Schrödinger. Models are specified by their "determining parts," or state variables, together with rules that relate these variables at one instant to their values at later instants. (Models frequently also have free parameters which must be measured and possibly adjustable constants.) The state variables, which fundamentally characterize the model, are discovered partly from experience and partly by exercise of the imagination. (For example, in Newton's model of the solar system, the state variables were the positions and velocities of the planets, the adjustable constants were the masses of the planets and the Sun, and one free parameter fixed the strength of gravity.) The model is inevitably far richer in details than any set of observations possibly could be. But having constructed one's model, one forthrightly attributes its elements to Nature and, without "dictating from the writing table" which measurements are possible and which are not, begs the experimentalist to test some of its predictions. (Schrödinger has just expounded Boltzmannism. Naïve realism he seems to disparage later in the article, but drawing the distinction is never easy.)

Models may be good or bad, may have too many variables or too few, but science without model-building is unthinkable. That is Schrödinger's first point. The strangest aspect of quantum theory, Schrödinger remarks next, is the way it carries over the Newtonian list of determining parts—position, momentum, time, energy—unchanged, yet denies the reality of the Newtonian *state*. "The classical model plays a Protean role in quantum mechanics," Schrödinger observes. Equally strange, one can know at most *half* the classically required parts, while the other half fades into "indeterminacy."

Schrödinger then asks: what is the meaning of this fade into indeterminacy? It cannot be the mere recourse to classical probabilities, since many classical formulas are either false or absurd in quantum mechanics. For example, in quantum mechanics a particle in a force field has a small but nonzero chance of being found where it could not go classically—because its kinetic energy would be negative and its velocity imaginary. Quantum theory deals with such conundrums by denying that the classical variables are simultaneously knowable, but "does one not get the feeling that the essential contents of what is being said can only with some difficulty be forced into the Spanish boot" of statistical predictions? (Schrödinger suffers from a blind spot here: some classical quantities might lose their meaning in an alternative interpretation; e.g., velocity in Féynes' model. Such imaginative failures were common in these early discussions; recall von Neumann's inability to imagine a model like Bohm's.)

If classical probability is out, what are the other possibilities? Perhaps the classical variables have become "washed out" or "blurred." An alpha particle resident in a radioactive nucleus might be smeared about, since inside the nucleus, remarks Schrödinger, "blurring doesn't bother us." But when the alpha particle emerges and hits a fluorescent screen, it is no longer blurry; the screen does not glow all over, but emits a miniscule flash.

"One can even set up quite ridiculous cases," says Schrödinger, prefacing the most famous (perhaps I should say infamous) of all quantum thought experiments. He proposes a cat locked in a steel chamber with "the following diabolical device (which must be secured from direct interference by the cat)": a vial of cyanide, a hammer to smash the vial, and a Geiger counter whose activation trips the hammer and kills the cat. If the Geiger counter is equipped with a bit of radioactive material, whether it records a "count" or not will be determined by one or a few radioactive nuclei, which in turn are governed by quantum mechanics. So, if the state of the nuclei are "smeared out" or "blurred," so is the condition of the cat: it is smeared out between life and death, a ridiculous proposition.

Schrödinger is not advancing some wonderful new ontological conception here, as certain commentators have imagined. He is knocking down a straw man. His point is that, due to the possibility of amplifying small effects, any "smearing out" could jump straight up from the scale of electrons and protons to that of cats and people. It is not a conceivable interpretation of the quantum-mechanical wave function. (To my knowledge, no respectable physicist ever claimed it was. Schrödinger is just being thorough in exposing the possibilities.)

I return in later chapters to Schrödinger's conclusions and his criticism of the "deliberate about-face" in epistemology taken by Bohr and

Heisenberg. I note here that Schrödinger never accepted the Copenhagen interpretation of quantum mechanics.

UNCERTAINTY OVER THE UNCERTAINTY PRINCIPLE

Ask any scientifically literate person to name a surprising *conceptual* discovery made in the 20th century by a physicist, and she or he will likely mention Heisenberg's uncertainty principle. (Relativity is the second most common choice, while complementarity—although a far more radical proposal—is well down the list, in my experience.) Heisenberg's laws of uncertainty seem to have metaphysical, even moral, significance: it is as though Nature, by enforcing Heisenberg's rules, is chiding the human race for imagining that it could attain unlimited knowledge. One is reminded of that hackneyed line from the science-fiction films, first spoken by Violet Kemble Cooper to Boris Karloff (playing a deranged scientist) in "The Invisible Ray" (1936): "There are some things Man is not meant to know."

In previous chapters we learned that the uncertainty principles are surprisingly robust, easily shaking off naïve attempts to falsify them. We also noticed that they display a sort of semantic uncertainty (imprecision? latitude?) in formulation and application. So it is time to bluntly ask: what do these laws actually say?

Unfortunately, the experts have not agreed on the answer. Dozens, perhaps hundreds, of interpretations of Heisenberg's laws have been proposed. The situation has become quite muddled. In hopes of clearing the waters somewhat, I begin by making an arbitrary assignment of terms to concepts. "Uncertainty" will henceforth refer to statements about you or me—that is, about our (or the observer's) state of knowledge or ignorance. "Indeterminism" will refer to some lack of definiteness on the part of the particle (or on the part of Nature, if you prefer). "Precision" or "latitude" will mean the smallest scale for which measuring apparatus can detect any variation. (All devices, of course, possess such limitations; for the measurement of position, the precision is given simply by the smallest perceptible division on the meterstick.) Finally, if randomness should make an appearance, let the phrase "statistical spread" refer to a measure of the amount of scatter in the data taken in a long run of identical experiments. (For those who suffered through a statistics course in college, the sample standard deviation will do nicely.)

In the following I will not be completely precise, omitting numeric factors and other details, and I will restrict the discussion to Heisenberg's position-momentum uncertainty relation. (The time-energy one is even less well understood.) As convenient labels for the two quantities, I adopt the physicist's traditional letters: "Q" for position and "P" for momentum (permitting the physics professor to admonish his students to "mind your P's and Q's.") I indicate the degree of precision with which we can measure a quantity with the traditional Greek lowercase letter delta; thus δQ is the minimum possible latitude left after a measurement of Q. I use uppercase Greek delta for the statistical spread: ΔQ will stand for the statistical spread in Q. Here are the most plausible interpretations of Heisenberg's first law, beginning, for dramatic purposes, with the one usually found in popular accounts:

First Interpretation. It is impossible to know the simultaneous values of the position and momentum of a particle except as permitted by Heisenberg's law: the product of my uncertainty about Q and my uncertainty about P is always greater than Planck's constant h.

Comment. Due to the possibility of retrodiction, this is literally false, as demonstrated in "EPR." Let us define knowledge to be "useful" if it can be used in a prediction concerning events at later times. Then here is a better version of the first interpretation:

First Interpretation (Revised Version). It is impossible to obtain *useful* knowledge of the simultaneous values of Q and P except as permitted by Heisenberg's law.

Comment. Since this interpretation seems closest to what Bohr and Heisenberg had in mind, I will call it the "party line." The meaning of the requisite uncertainty is not pinned down, but this is consistent with Bohr and Heisenberg's usage: sometimes it meant latitude and at other times statistical spread.

According to the party line, the important point is that one cannot get useful knowledge violating Heisenberg's rule. But even in Boltzmann's time, everyone knew that in a gas of 10^{24} molecules one could not get useful knowledge of the position or velocity of molecule number 176452905770084756394 78 in any event. To conclude from this that realism must be abandoned is a logical howler.

Second Interpretation. Quantities such as P and Q are indeterminate in the microscopic world of atoms, except as permitted by Heisenberg's law: the product of the indeterminacy in Q and the indeterminacy in P is always greater than h.

Comment. This interpretation puts the blame on the particle but has three problems. First, it can only be justified by statements about what one can know, so it is little more than a prejudice. Second, it is vague: what does it mean to be "indeterminate"? (Not "blurred"—remember the cat!) Third, due to the possibility of retrodiction, it cannot be demonstrated. In fact, it is *false* unless you rule out rectilinear classical motion by fiat, which would be circular logic.

In short, this interpretation cannot be defended.

Third Interpretation. There is no intrinsic limitation on the accuracy with which we can simultaneously measure the position and momentum of a particle. However, we do not always get the same values for Q and P, even if we set up the experiment ("prepare the state of the particle") and carry out the measurement in the same way every time. These quantities *fluctuate* from one run of the experiment to another. The statistical scatter in the data from a long run of experiments obeys the law

$$(\Delta Q) \times (\Delta P) \geq h.$$

Comment. In this interpretation, position or momentum changes unpredictably, like the velocity of a feather as it glides slowly to the floor. This randomness might be pure stochasticity (God throwing dice) or deterministic "chaos," but neither vagueness on the part of Nature nor human limitations are implied (except with respect to predicting the future). I am tempted to call it the "realist" interpretation, since the heretics Bohm, Fényes, and Nelson would likely support it. (This is certainly the case for Fényes, who wrote: "the fact that the Heisenberg relation holds in no way excludes the simultaneous measurement of position and velocity... to arbitrary accuracy." Bohm also considered it possible, in principle, to measure both quantities with "unlimited precision," but his views are best included under the next interpretation.) Since the next proposal can also be called "realist," I will call it the statistical interpretation. Heisenberg and Bohr would have been horrified by this notion, of course, as it reduces the status of the uncertainty laws from metaphysical prohibitions to more familiar statements of the human condition; we are equally uncertain of the outcome of a coin toss.

Fourth Interpretation. The key to the uncertainty principle lies in the *measuring apparatus*. If we first measure Q with latitude δQ (the precision being selected, let us say, by adjusting a dial on our apparatus) and then make a second measurement of P *on the same particle* with great accuracy, we will find a large statistical spread ΔP in P (with a symmetrical statement for δP and Q). These reciprocal relations can be expressed as:

$$(\delta Q) \times (\Delta P) \geq h,$$

$$(\delta P) \times (\Delta Q) \geq h.$$

In other words, the smaller the latitude in the measurement of the first quantity, the greater the statistical spread in the second.

Comment. This interpretation invokes an "uncontrollable perturbation caused by the act of measurement," UPCAM for short. Physicists, who have as much trouble as the rest of us mastering Bohr's complementary thought, quote UPCAM if pressed. (An example is Feynman; see below.) UPCAM simply claims that a measurement of one quantity produces perturbations in the other, which are random and with magnitude inversely proportional to the level of precision selected. No agnosticism is implied about the particle "possessing" either quantity before or after the measurement. Therefore, this is also a "realistic" interpretation. (Conceivably, the quantities are "possessed" only by particle-plus-instrument, being created by their interaction; see Section 6 ahead and Chapter 18,

"Principles." Provided the details are given, this could also amount to a realistic interpretation.)

Since the statistical spread logically cannot be smaller than the precision, the relation $\Delta Q \times \Delta P \geq h$ of the last interpretation follows from the relations above. Thus it is only a refinement of that interpretation rather than a disjunction. But physically it is very different, since it pins down the causes of the quantum fluctuations: they are in the measuring instrument.

Although UPCAM is quite popular, one rarely encounters a person who claims to know *what kind* of perturbations are involved, or why they are uncontrollable. (When I ask this question, most physicists just shrug, although many make a mental note to avoid my company in the future.) An exception was David Bohm. In the second of Bohm's 1952 papers, there is an explanation of the scatter in P that would be produced by a precise measurement of Q on the basis of violent fluctuations in the "quantum force" acting on the particle. Another explanation invented by Bohm and a colleague is discussed in Chapter 18, "Principles."

There are, of course, many more interpretations in the literature. Which one—the party line, statistical, UPCAM, or some other—is the "right" one? I cannot say. Most theorists, who never make use of the principle in calculations, are disinterested. Nor are many experimentalists trying to make as precise measurements as possible on an elementary particle and thus demonstrate the workings of "uncertainty"—although there are some. The main function of Heisenberg's laws, like Bohr's complementarity interpretation, is psychological: it lets one stop worrying that the real picture has been missed. (Einstein once compared the "Bohr-Heisenberg tranquillizing philosophy" to a soft pillow on which to rest one's head.) But I have a hunch that a better understanding of quantum uncertainty might help us grasp what is going on "down there."

QUANTUM PROBABILITY, OR WHAT PHYSICISTS DO NOT SPEAK ABOUT

What does it mean to say that something is "random"? A lot of ink has been spilled over this issue since the 17th century. There is the objective stance: a coin is random because of its observed behavior in a long run of flips. There is the subjective view: the coin is not random; the randomness lies in my unsettled opinion. At least five other theories are known to me. But in this section I must refer to all of these interpretations as "classical." For, at least in the opinion of the Nobel Prize–winning physicist Richard Feynman, quantum probability is just not the same as classical probability.

In 1951 (note the date), Feynman gave a lecture to an audience of mathematicians and statisticians in Berkeley on the subject of probability in quantum mechanics. I read the resulting paper 20 years later; it was one of the principal sources of my personal angst about quantum theory.

Feynman begins with an account of the two-slit experiment and how quantum mechanicians treat waves as probabilities. This approach violates the rules for adding probabilities of mutually exclusive events going back to Laplace, says Feynman. "Many... classical mechanisms have been tried to explain the [two-slit pattern], but none of them in the end has proved successful." (The very next year, Bohm will invent one!) Feynman next remarks that if each particle passes through just one of the two slits, the pattern formed on the screen would be different. This yields a paradox. He considers various possibilities for escaping from the paradox but dismisses them all. For instance, one might imagine that the electron passes through both slits, perhaps going first through one slit, then out again and through the second, etc. But by shining light near the slits, one can actually determine which one the electron passed through. But if you do so, says Feynman, UPCAM changes the pattern to that of two one-slit patterns superimposed. "We thus see that any physical agency designed to determine through which hole the electron passes must produce, lest there be a paradox, enough disturbance to alter the distribution from [the two-slit to the one-slit pattern]."

Feynman then reveals how physicists treat these logical difficulties:

> We are still left with the question, "Do the electrons have to go through hole 1 or hole 2 or don't they?" To avoid the logical inconsistencies into which it is so easy to stumble, the physicist takes the following view. When no attempt is made to determine through which hole the electron passes, one cannot say it passes through one hole or the other. Only in a situation where an apparatus is operating to determine which hole the

> electron goes through is it permissible to say that it passes through one or the other. When you watch you find that it goes either through one or the other hole, but if you are not looking you cannot say that it goes one way or the other! Such is the logical tightrope on which Nature demands that we walk if we wish to describe her.

I would love to have seen the expressions on the faces of those lovers of logic at the end of Feynman's talk. Here, said the brash young physicist, is a predicament that you, with all your powers of mathematical analysis, cannot understand. But we physicists are not troubled; we simply refuse to speak about the situation. (Perhaps John Bell reflected on Feynman's words when he chose as the title of his collected papers: "Speakable and Unspeakable in Quantum Mechanics.")

Very well, some things will not be spoken of. But why did Feynman chose as his topic *probability* in quantum mechanics, rather than unspeakability? Perhaps Feynman had another puzzle on his mind that day in Berkeley. "When I say that the probability of an outcome in an experiment is p, I mean the conventional thing," Feynman reassures his audience at one point; ". . . no departure from the concept used in classical statistics is required."

Classical statistics was invented to help experimentalists distinguish between types of variability in their data. For purposes of discussion, I here lump such variability into two categories: measurement errors and "real fluctuations." In clinical trials doctors often give a drug and then monitor some interesting marker, such as blood pressure or numbers of circulating white cells. The outcome changes from patient to patient, and usually a statistician is called in to analyze the data and determine what fraction of the changes is due to errors in measuring the pressure or counting the cells, and what is due to true variability in the patient's responses. Pharmacologists, geneticists, and biologists then try to discover the causes of the residual "real changes."

Quantum mechanics, as far as I know, is the first theory in the history of Western science to predict randomness but to deny that *either* measurement errors *or* "real fluctuations" account for them. It is a curious situation. The orthodox Copenhagen-er of course rejects both interpretations, taking the unanalyzability of the measurement process as cardinal doctrine. Other physicists attribute randomness to UPCAM but think it absurd to scrutinize the perturbation the way a medical scientist might investigate how a drug alters a patient's metabolism. That would mean returning to those old-fashioned, "mechanical" models of atoms (or of measurement devices), which is too philosophically retrograde to be considered.

Perhaps this is what was on Feynman's mind when he chose to lecture probabilists and statisticians about chance. (Not to mention the pleasure of puzzling a group of which Feynman had a very low opinion.) What should this novel form of randomness be called? Unfortunately, Feynman did not speak of that.

TIME AND THE STATISTICS OF PARTICLES AND PEOPLE

Quantum mechanics does a wonderful job of predicting where an electron is likely to be found at a given time. Just look where Schrödinger's wave function has a large amplitude, it says; where the amplitude is small, the particle will be scarce. Knowing this and a little bit extra (namely Pauli's exclusion principle forbidding two electrons from occupying the same state), one can derive the spectrum of the hydrogen atom, the qualitative features of the periodic table of elements, the strength of chemical bonds, and much, much more. This was a splendid achievement.

But quantum mechanics is less admirable when it comes to predicting where the electron will be found if a second measurement is made at a later time. In fact, the theory seems to be ambiguous on this point. Although this problem might be solved by fiat—Memo from the director of the physics laboratory: *Given the present situation in quantum theory, until further notice no measuring equipment may be connected in sequential arrangements*—it is natural to repeat measurements. But there are also simple, tractable experiments that cannot be carried out entirely at a preselected time. Consider the following: at 12:00 noon I open a door on an atomic oven containing molecular hydrogen. One meter away, and connected by an evacuated pipe, is a hydrogen detector. The measurement just consists of noting the first time the detector detects a hydrogen molecule.

Conventional quantum mechanics, faced with this trivial situation, is strangely helpless. The reason is that the statistics of arrival times cannot be inferred from any amount of purely positional information. Let me illustrate this with a little story, which I hope will not sound sexist. A woman frequently asks her husband to go to the bank to cash a check, always making the request a few minutes before 4:00 P.M. (The bank closes its doors promptly at 4:55.) He always returns a little after 5:00, but usually with less cash than he should have, and sometimes with none at all. She has a good guess as to the explanation: he is always gone a full hour and spends most of the time—it only takes five minutes to cash a check—at the neighborhood bar, which happens to be next door to the bank. But what (she wonders) is his routine? Does he go straight to the bar and then to the bank, or the reverse?

To investigate the matter, she decides to call the bar at randomly chosen times between 4:00 and 5:00 (only one call on a given day) and gather some statistics on the errant husband. The data from her experiment reveal a simple situation: regardless of when she calls, the husband is at the bar 11 times out of 12. That is, for 1 call out of 12 he is missing. Question: can she deduce from this information something about the time at which her husband arrived at the bank—or whether he arrived at all?

Answer: no. There are alternative explanations.

First Scenario. Her husband always goes directly to the bar. At some random moment between 4:00 and 4:55, he departs for the bank, returning to the bar if there is time remaining in the hour. Since he is absent for exactly 5 minutes out of 60, the probability of his being absent at the moment she calls is 5/60, or 1/12.

Second Scenario. On different days he does different things. On most days (exactly 11 out of 12, in fact) he goes directly to the bar, starts drinking, and forgets to go to the bank at all. One time out of 12 he goes to the bank, cashes the check, and then falls into conversation with the bank clerk (who happens to be his ex-girlfriend) for the entire hour.

There are, of course, other explanations. How might our puzzled wife discover the real situation—assuming checking his wallet or smelling his breath to be cheating? By making two calls to the bar, of course, separated by more than five minutes. In the second scenario, in 1 case out of 12 he will still be absent on the second call. The problem with the original data was its incompleteness: although after many calls the wife had a *complete knowledge of the husband's spatial probability distribution*—namely, with probability 1/12 he was at the bank and with probability 11/12 he was at the bar—*for all times between 4:00 and 5:00*, she could not deduce anything about when he reached the bank, or whether he got there at all. Spatial information about random events cannot tell the whole story.

One could argue that removing this defect was precisely the goal of the heretical theories in which particles have paths. (I suspect that it was also Feynman's motivation in introducing his "sum over histories," or "path integral," approach to quantum mechanics in 1951. Public orthodoxy may conceal private doubts.) Both Bohm's and Féynes'/Nelson's model solved the problem of random events such as arrival times, but they do not necessarily give the same answer. This is revealed by imagining a strange beast called "π-mesic hydrogen."

In π-mesic hydrogen the orbiting electron is replaced by an exotic negatively charged particle called a π-meson (or simply "pion"). This peculiar "atom" behaves like hydrogen, except that the greater mass of the pion changes the energy levels, and one other fact. The pion belongs to a different family of elementary particles than the electron and can undergo a nuclear reaction with the proton in which the two annihilate each other, leaving behind a neutron and some gamma rays (photons). This only occurs when the pion gets close enough to the proton to feel the extremely short-ranged "strong force" that holds nuclei together.

What do Bohm's model, the Fényes/Nelson stochastic mechanics, and quantum mechanics say about the stability of pionic hydrogen? In Bohm's model, in the lowest-energy stationary state the pion simply sits at rest in space—so in his theory π-mesic hydrogen appears to be stable, although some fraction of the atoms might undergo spontaneous annihilation the instant they formed. In the Fényes/Nelson theory, the pion moves around endlessly like a pollen grain in water and so is sure to eventually wander too close to the proton and cause a reaction. E. Carlen (an ex-student of Nelson) of MIT and A. Truman of University College, Swansea, Wales, have calculated this probability and proposed observations of π-mesic hydrogen as a test of stochastic mechanics. As for conventional quantum mechanics, it can calculate average reaction rates, as Arthur Wightman of Princeton showed for negative muons absorbed in hydrogen in the 1940s; but to calculate the full time-distribution in this pathless theory would likely require detailed knowledge of the strong force, which now exists as part of the so-called standard model known as quantum chromodynamics, as well as a complete solution to the measurement problem. The calculations might be intractable—and, as we shall see below, the answer may depend on whether or not anyone is watching this strange atom.

In quantum mechanics, particles spontaneously jump from state to state and are emitted, absorbed, or annihilated; yet the times of these exotic events are not given in the theory. Some temporal denial has been going on in physics since 1925.

CAN A GRADUATE STUDENT COLLAPSE THE WAVE PACKET?

As we saw in the opening section of this chapter, sometimes one must collapse the wave function. The problem is when this should be done, and why. I leave "when" to the last section and deal here with "why." Physicists have some very peculiar opinions about this issue.

Complex numbers cannot represent measurements directly. So Max Born added a rule, called the "projection postulate," for getting the data out and simultaneously updating the state. Consider measuring the energy of a solitary hydrogen atom. Hydrogen has many stable states of definite energy, each represented by one of Schrödinger's proper vibrations. But the atom's wave function need not be equal to any one of these; it can be a sum (called a "superposition") of two or more, each entering in with certain "weights." Suppose the wave function is the sum of two waves, the first with energy of one unit and amplitude multiplied by

3/5 (the weighting factor); while the second has energy two units and is multiplied by 4/5. (The choice of the fractions is arbitrary, except that their squares must add to one.) The physicist writes this as

$$\frac{3}{5}\psi_1 + \frac{4}{5}\psi_2,$$

the Greek letter ψ being Schrödinger's choice to represent his wave function, a choice which has stuck. Then Born's prescription is that, in a measurement of the energy, with probability 9/25 one gets one energy unit and the state collapses instantly to the first wave (ψ_1), and with probability 16/25 one gets two units and the state collapses instantly to the second wave (ψ_2).

It is simple enough—but what is responsible for this instantaneous, discontinuous transformation of the state of the system? Schrödinger's dynamical law is not the culprit, because it makes the wave evolve continuously. It cannot be a simple updating of knowledge, because the measurement changes more than just our ignorance of the thing measured. Indeed, where quantities related by "reciprocal uncertainty" are concerned—position and momentum, or time and energy—collapsing the state to a sharp value of one results in one's knowing *less*, not more, about the other.

Ignoring philosophical quick-fixes, only the measuring apparatus can come to our rescue. The question of how this might work is called the Measurement Problem and has generated a huge literature since the search for a solution was initiated by von Neumann in 1932. John Bell argued in his provocative last paper "Against 'measurement'" that all attempts to solve the measurement problem in standard quantum theory are doomed to fail. (See Chapter 18, "Principles.") But the orthodox must nevertheless try; the attempts thus far have generated some remarkable speculations.

As Feynman remarked in his 1951 article, "The world cannot be half quantum-mechanical, half classical." (With this remark Feynman uttered a heresy. Feynman was not a realist as I understand the term, but this statement proves he was not a Bohrian dualist, either.) Apparatus are made of atoms; so, Bohr's philosophy aside, it must be possible to describe the measuring apparatus quantum-mechanically. How might this work, mathematically? A particle has an associated Hilbert space (as described in Chapter 8), the fictitious "space" of all wave functions, any one of which might describe the "state of the particle." The apparatus also has a Hilbert space, but these waves live on the enormous configuration space of the apparatus, which must include enough numbers to specify the location in space of a trillion times a trillion atoms, with still more for each atom's internal configuration. The Hilbert space of the combined system is even larger, consisting of all wave functions on the configuration of particle-plus-apparatus.

Now we can let the wave function of the whole thing evolve according to Schrödinger's law. This satisfies the demand for a unified treatment of the macroscopic and the microscopic, but have we made any progress? Apparently not. Von Neumann showed in his book how to construct situations in which, after the particle and apparatus have interacted for a time, the state in the big Hilbert space is a sum of states corresponding to "pointer on the apparatus reads one energy unit and the particle has state ψ_1" (and so forth), each of which contains a macroscopic and therefore potentially observable part. But the overall state is still a complex wave function stretching over all these possibilities. Neither real data nor randomness has yet materialized.

Then let us add another apparatus to "read" the dials on the first apparatus. But now we detect the infinite regress involved: soon enough we shall arrive at the wave function of the whole universe. At some stage, an outside agency is going to have to come in and collapse the wave function in one of these cumbersome Hilbert spaces.

Inevitably, the idea arose that the consciousness of the observer might terminate the process and resolve the paradox. The wave function might be collapsed at the first instant that someone perceives the result of the measurement. Note that in this proposal the mind of the observer acts directly on the particle, and that "mind" cannot be replaced by "brain," since the latter is a merely a cauliflower-shaped mass of neural, vascular, and connective tissues, yet another system to which quantum-mechanical laws ultimately apply. No, in order to play its transcendental role, a nonmaterial agency must be invoked; mind fills the bill. (I need not point out the joy with which the parapsychologists greeted this proposal. Theists, of course, made yet another suggestion.)

Some amusing speculations suggest themselves. Does the collapse take place when the experimentalist first glances at the dials on the apparatus, when she writes the resulting paper, or only at the time of its publication? And who is competent to do the deed? As Bell asked in "Against 'measurement',"

> What exactly qualifies some physical systems to play the role of "measurer?" Was the wavefunction of the world waiting to jump for thousands of millions of years until a single-celled living creature appeared? Or did it have to wait a little longer, for some better-qualified system ... with a Ph.D.?

Then there is the paradox proposed by Eugene Wigner (von Neumann's friend and Nobel Prize–winning quantum theorist): suppose you first have a friend read the dials on the apparatus, then ask him about the result. Since your friend is another material system, the hegemony of quantum mechanics over everything requires that his brain state, now correlated with the apparatus state, which is correlated with the particle's state, also be represented in the gigantic Hilbert space of the total system. Therefore, your friend's brain state is a superposition of possibilities, like Schrödinger's cat before we open the box. Since one cannot believe that one's own mental state can be so unsettled, it follows that the friend cannot really be conscious. Therefore, no one is truly conscious other than you, or me, or perhaps E. Wigner. The belief in the completeness of quantum mechanics has driven us to the most absurd philosophical position imaginable: pure solipsism.

Other solutions to the paradox of wave-packet collapse have been proposed within orthodox theory, but all are too absurd to believe. Somewhere between the microscopic world governed by complex numbers and the macroscopic world governed by real ones, a signpost must exist reading "quantum mechanics ends here." But no one knows where this sign appears, or what happens at the "infamous boundary" between two worlds.

"A WATCHED POT NEVER BOILS"

Recently I made some experiments to test this old adage. Pasta water, as measured with an egg-timer, boils at the same rate watched or not, I found. However, in another series of experiments, I observed that immature humans exhibit an "observer effect." Five-year-olds, for example, make fewer illicit raids on the fridge if watched. Children, therefore, are non-classical, like electrons—a cryptic remark, to be sure, but one that leads to perhaps the strangest paradox of quantum mechanics.

Consider the following experiment. An electron is emitted at 12:00 P.M. into a vessel whose walls contain many small holes. Eventually the electron will find its way out, perhaps first bouncing many times off the walls. In order to record this event, the vessel is completely surrounded with electron detectors, which detect (and absorb) the electron if it succeeds in escaping. What is the probability that at 1:00 P.M. the electron is still in the box?

Now consider a second experiment. Rather than relying on our bank of detectors, we open the box, shine a light inside, and take a quick peek. (Never mind that we would need gamma rays, as in Heisenberg's microscope.) If the electron is still in the box, we instantly close it up, wait a moment, and repeat the procedure. To assure a parallel situation, the box is surrounded with perfect "electron absorbers." Same question for this experiment.

Classically, these probabilities would seem to be equal, but are they also so in quantum mechanics? Curiously, there is no agreement among physicists on this point, as an amusing episode I recount below makes clear. But first, I contrast the classical versus the quantum reasoning about this situation.

Suppose that at 12:05 we shine light into the box and observe the electron in the lower-left quadrant. How can we make use of this information? A classical realist would argue as follows:

"At noon, if we know nothing more than that the electron is in the box, we might assume that it is equally likely to be anywhere inside. After all, probability is only a numerical measure of our ignorance of the electron's position. But with the new information, we must "collapse our ignorance" to say that the electron is now (at 12:05) somewhere in the lower-left quadrant. ("Collapse" here means replacement of probabilities by conditional ones, as explained in Chapter 12.) As we take more peeks, these benign "collapses" tells us more and more about the particle's trajectory on this one occasion. But over many experiments, we find the same rate of escaping as if we didn't watch."

In other words, a watched (classical) electron escapes if it would.

By contrast, most quantum physicists would reason like this:

"At the start of the experiment, the wave packet representing the electron is contained entirely in the box. But as time goes on it 'leaks' out. In the second scenario, each time one peeks and sees the electron, one has to collapse the wave to be back in the box. If we increase the number of these peeks and collapses during the hour, the wave becomes 'frozen' in place. Consequently, the probability the electron remains in the box for the whole hour approaches one!"

In other words: a watched (quantum) electron never escapes. (Mathematically, what happens is that, unlike classical probability, quantum probability can change only by an amount proportional to the time interval squared, i.e., in 0.01 sec the former changes by (say) $3 \times 0.01 = 0.03$, while the latter might change by $3 \times (.01)^2 = .0003$.)

So where is the paradox? Microscopic events simply obey different rules than macroscopic ones. Somehow directly observing (and, presumably, perturbing) the electron slows it down—although one might have thought UPCAM would *speed up* a particle. But it is not that simple. There are two troublesome issues raised by the "watched-pot-never-boils" effect in quantum mechanics.

The first is a question of principle and can be stated in the following overly dramatic way: is it true that a watched cat never dies? Recall Schrödinger's unhappy feline locked up with a radioactive nucleus, a Geiger counter, and a vial of poison. Let there be a clear glass window in the wall of the cat's prison. You sit outside and stare fixedly at the captive. Since puss is safe until the alpha particle emerges and triggers the Geiger counter, observing her amounts to measuring the nucleus' state. But since you are observing continuously, wave collapse is going on continuously; so the nucleus never decays, and puss lives on—at least until you lose interest.

The principle under discussion is Born's projection postulate. Some textbooks declare that one projects (or collapses) the wave function each time one learns something new about the system. The immortalized cat reduces this to absurdity, because if we do not interact with the microsystem directly, a disturbance cannot be invoked to resolve the paradox. (It is hard to believe that the photons reflected off the cat's fur give her, or the radioactive nucleus, a serious case of UPCAM.) In less dramatic fashion, in the first version of the leaky-box experiment it is absurd to suggest that the state of the particle *while it is still in the box* changes every time I glance at the detectors.

The second question is Bell's query: how large does a system have to be before it can carry out a "measurement"? Its essence is perfectly captured by an episode from 1983, in which two physicists seriously proposed, in the pages of America's leading physics journal, that protons in the nucleus are stable only because they are under continuous observation by the neutrons.

This episode requires a bit of background information. Until sometime in the 1970s the proton was thought to be stable: left to itself, it never decays spontaneously into other particles. But then came the "grand unified" theories of elementary particles, quantum theories that try to unify all known forces. One variety of grand unified theory predicted that the proton was unstable, but only on a very long time scale: typically, 10^{31} years would pass before a single proton would decay. (This is an average figure; some protons would take less time to decay, some more.) One might take this to mean that proton decay will never be observed—the whole history of the universe from the Big Bang to the Final Crunch may take only 10^{15} years—but this is not the case. If one instruments a large block of material, say a cubic kilometer of sea water or a hundred tons of liquid gallium, one can keep more than 10^{31} protons under observation, in the sense that one can watch the whole volume for the tell-tale "flash" of particles revealing that a proton has decayed. In a year's time one should see a few decays. Many teams of experimentalists have raced since the early 1980s to be the first to catch a proton in the act of decaying; a Nobel Prize surely awaits the winner, if there is one.

After several years of "null" outcomes, two physicists, L. P. Horwitz and E. Katznelson of Tel Aviv University, suggested in *The Physical Review* that the watched-pot-never-boils effect explains the proton's stubborn refusal to decay in these experiments. The protons in question reside not in empty space, Horwitz and Katznelson pointed out, but in the heart of the atom, the nucleus, imbedded in a "sea" of neutrons. Each 10^{-24} second (one trillionth of a trillionth of a second) on average, a neutron encounters the proton, bounces off it, and thus can be considered to have "measured" its continued existence. Alternatively, since the information about the proton's absence would be spread very rapidly through the nucleus by collisions among the other nucleons, one might consider the whole nucleus to represent a "proton measuring device." (If one considers a proton to be a little bag of quarks, as most physicists do, then decay of a proton means that certain particles produced by quark decay shoot out of the bag. Thus our neutrons play the role of the photons in the second *Gedanken* experiment.) Thus, every 10^{-24} second, the wave function must be collapsed, and this increases the proton's lifetime enough that catching one decaying really is unlikely.

Now for the denouement of this exciting episode. Six months later, *Phys. Rev.* published three letters "refuting" the Israeli physicists' provocative theory. In each case the writer(s) attacked the reasoning of the authors, repudiating their use of wave-packet collapse—but, oddly, not for the same reasons. Three physicists from Cornell simply denied that one collapses the wave packet, with no discussion of when one does or why not in this case. Two physicists from Oxford (one was John Bell's thesis advisor, Rudolf Peierls) made the same claim but at least mentioned the measurement problem, remarking that collisions within the nucleus "can become an element in an observation if the collision is followed by an observation...," but then left the issue dangling. A physicist from New Mexico proposed that one only collapses the wave if a particle comes out of an experiment and goes behind a baffle. (Should I call this idea baffling? At least it is an explicit suggestion for resolving the paradox.)

The watched-pot-never-boils effect, also called the quantum Zeno effect, was experimentally verified, perhaps for the first time, in 1989. (One doubts that Zeno of Elea (336?–264? B.C.) conceived of wave-packet collapse, although he gave other reasons why objects might not move at all.) The beautiful experiment demonstrating the effect was performed by J. Bollinger, D. Heinzen, W. Itano, and D. Wineland at the National Institute of Standards (previously the National Bureau of Standards) in Boulder, Colorado. The "water" in their experiment consisted of 5000 beryllium ions trapped by magnetic fields. During a time interval of 256 seconds, a radio wave drove some of the ions from a ground state to an excited state of higher energy (representing "boiling"); call it level two. While this was going on, the experimenters occasionally sent in a brief pulse of laser light, which can drive ions up to a level three. An ion can only make the transition to the third level if it is in the ground state when the laser pulse arrives; it then drops quickly back, emitting a photon of the same color as the laser. Thus the amount of light coming out of the trap after a pulse is sent in reveals the fraction of ions still in the ground state.

When the experimenters "peeked" with the laser after a 256-second interval, essentially all the ions had jumped to level two. If they peeked twice, once at 128 seconds and once at 256 seconds, only about half made the jump; if they looked four times, the figure dropped to a third. At 64 peeks, almost all the atoms stayed in the ground state. The water never boiled.

17.

Philosophies

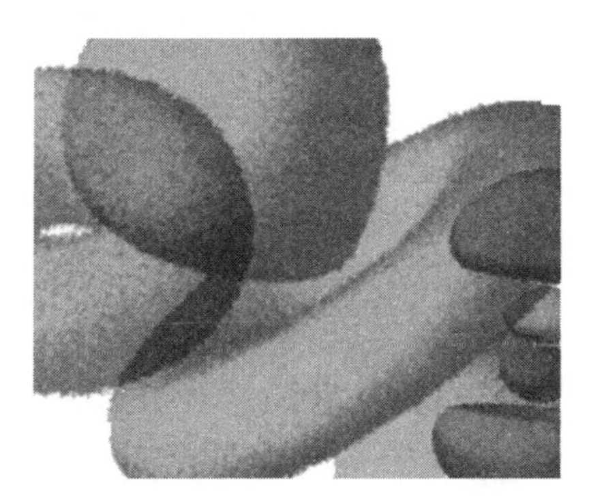

Scientists sometimes pretend that what they do is as natural as breathing, and hence they have no philosophies. But Heisenberg, Bohr, Schrödinger, and Einstein all paid attention to theirs, and indeed no scientific discipline in this century has attracted more philosophical comment than quantum physics. (Evolutionary biology comes close in this department, but by now probably lags behind.) Should one not therefore make a compendium of these views, the better to sift truth from error?

Unfortunately, the task is comparable to listing the concepts of god in the world's religions, or the loopholes in the Tax Code. Max Jammer, a philosopher and historian of science at Bar-Ilan University, did carry out a survey in 1974; his excellent book *The Philosophy of Quantum Mechanics* reported on (if his name index may serve as a yardstick) the opinions of about 500 scientists and philosophers. In the meantime the deluge has not abated. Besides the sheer volume of this material, it is of highly variable quality. Reading it is like walking barefoot through a thistle-filled field searching for an exotic flower: you can only do it for so many hours per month.

I therefore confine the discussion to the positions of the historically most prominent players. Their philosophies have been described to some extent in previous chapters; the goal now will be to acquire some more general points of view, which will help us navigate over the difficult terrain on which the battle was fought. (Identifying the *principles* of quantum theory, if possible at all, is the task of the next chapter.) Having placed these markers, locating Einstein's and Schrödinger's position on the map will not prove difficult, but Bohr's will require additional effort. As a philosophy may reveal its true nature when transplanted to a novel setting, I digress from physics at the end of the chapter to describe the adventures of Bohr and his followers in biology, and the influence the psychologist William James may have had on Bohr's formulation of the complementarity principle.

Bohr and Einstein debated many issues, but at the core of their disagreement was the same kernel that had divided Mach from Boltzmann: the proper attitude toward theories. At one extreme we have the position called, usually by its detractors, "naïve realism." (As virtually every scientist imagines himself or herself to be a realist, the qualifier is a practical necessity. Philosophers prefer other modifiers, e.g., "critical" or "representational"; see the Notes.) First and foremost, this represents a desire to describe a world of properties or qualities existing in space and time, independently of any observer. The naïve realist admits that a measurement might affect the thing measured, but if so he or she demands the details. For example, measuring the temperature of a drop of water with a standard medical thermometer will alter the drop's state, and give a misleading result to boot. But the large heat flow relative to the small heat capacity of the drop gives a satisfying explanation of the perturbation. In short, Berkeleian subjectivism is anathema in this philosophy.

Flowing naturally from naïve realism is a belief in the speculative freedom of the theorist, on the one hand, and the pragmatic freedom of the experimentalist, on the other. According to the first, the theorist is free to construct elaborate models containing ideal entities, e.g., "electrons," and attribute the details to reality. (This freedom is perhaps the essence of Boltzmannism.) The theorist encourages the experimentalist to measure whatever he can, unhindered by restraints "dictated from the writing table," as Schrödinger put it. If the experimentalist can measure little of what the theorist proposes, this must be accepted philosophically; one cannot expect observations to confirm more than a few consequences of any model. But just as the theorist may not place signposts reading "out of bounds to experimentalists" in her speculative creations, so experimental limitations place no *a priori* constraints on the theorist's imagination. (Experimental *results*, of course, yield *a fortiori* bounds on the exercise of that imagination.) Ideally, theorist and experimentalist are less master and slave than business partners: they prosper or go bankrupt together.

I hasten to point out that naïve realists may be naïve about their realism but not about how science works. Indeed, no one mentioned in this book would subscribe to the simplistic "observational-inductive" model of scientific progress popular among journalists and textbook writers. In this scenario the experimentalist arrives at her laboratory each morning uncontaminated by the opinions of theorists, observes the wonders of nature almost as a child, and passes her observations on to the theorist, who endeavors to "explain" them. This vision better fits a hunter-gatherer relating the day's events to an elderly shaman than an experimentalist at the controls of a particle accelerator sending e-mail to the theorist at her computer workstation.

As Einstein frequently remarked, it is theory that tells the experimenter what a measurement *is*—how hooking apparatus together in certain arrangements can amount to a measurement of something. (It may seem trivial, but reflect on the fact that experimentalists typically must construct apparatus before making observations. Stern-Gerlach analyzers, for example, do not self-assemble in the night.) All knowledgeable observers of science know it has a "hypothetico-validational" structure, in which theories, motivated partly by experimental facts and partly by imaginative hypotheses, gain increasing plausibility or are falsified by experiments. (Experiments cannot *prove* theories. *Mathematical theorems* can be proven, since they are assertions about formal systems, but in the factual world there are always alternative explanations.)

Naïve realism should be distinguished from "classical realism." Realism does not presuppose classical concepts. Classical physics described the world revealed by our senses, but a sensible object is made up of atoms, and what worked for the whole need not work for the parts. Macroscopic phenomena should be explained in terms of laws governing microscopic events, not the other way around. (Here we verge on the "holism-reductionism" question, another axis of contention I touch on below.) It would be surprising if these laws contained no elements foreign to classical thought.

Although agreeing that classical theories are dispensable, most naïve realists regard *understanding* as on a par with *prediction*. This is a sticking point, for "understanding" is inevitably subjective and history-dependent. It would be impossible to generate agreement among all naïve realists as to which features of a model remain in the arena of comprehensibility and which depart from it. Are fractal Brownian paths "understandable"? Is randomness?

The antithetical doctrine to naïve realism in physics is not, as a reader exposed to the popular literature might conjecture, "sophisticated mysticism." It is closer to what I call "simple instrumentalism." Most physicists retreat to this Machian position when pressed on the realism question. "The role of theory is to predict what we see on the dials of our apparatus," they say, "and if the predictions are accurate, the theory is good. Answering these other questions— about what is 'really going on'—is a meaningless exercise."

Bohr often expressed this sentiment in his writings. I quoted his approving description of quantum mechanics as nothing more than a probability catalog in Chapter 6. Here is another example: responding to EPR, Bohr wrote

> Such an argumentation, however, would hardly seem suited to affect the soundness of quantum-mechanical description, which is based on a coherent mathematical formalism covering automatically any procedure of measurement like that indicated.

This is pure Machism. If the equations work, says Bohr, the theory is beyond criticism.

The naïve realist rejects this view as absurd. Let me put the counterargument in the form of a parable:

One day, a group of physicists announced the discovery of the long-awaited "Theory of Everything," the ultimate theory that could predict in advance the outcome of any experiment that could ever be performed. The computations necessary for this feat could be carried out, the physicists argued, by a huge computer, which (they further proposed) might occupy one-third of the state of Alaska—the glacial ice there being useful to cool its gigantic bank of superconducting superchips. This superproject was going to be very, very expensive, but after an extended campaign, in which the physicists relied on their time-tested argument "If we don't do it, someone else will!" the project was funded and brought to completion.

Finally, the great moment arrived. The dignitaries, shivering in their parkas, had given their speeches and departed, and it was time to demonstrate the power of this ultimate computer. The first to pose a question was the director of the soon-to-be-obsolete Large Hadron Collider in Geneva, who typed in "I have a proton-proton colliding beam experiment with a center-of-mass collision energy of one-times-ten-to-the-twelfth-power electron volts, equipped with a Higgs boson detector, and I measure the flux into a solid angle of 2.73645383746453 sterradians," and in a flash the answer appeared on the monitor:

$$3.14159260467268 \times 10^5 \text{ cm}^{-2} \text{ sec}^{-1}.$$

The director admitted that this was correct, to all 14 decimal places. With an ashen face, she turned to go, and the next in line stepped up to ask his question. Another right answer. And so it went; experimental physicists studying everything from cosmic rays to gravitational waves to beach sand asked their question, received their answer, and departed. The theoreticians, it appeared, had finally solved the equation of the universe.

On another day, long after physicists had abandoned doing messy experiments, and the science called "experimental physics" had become a historical curiosity, a philosopher asked the leading theoretical physicist (whose job now consisted of entering data into the machine and making sure that it never ran out of ice),

"How does the program stored in the machine operate?"

"Um ... well, to tell you the truth, no one really knows," replied the physicist.

"What! How can that be? Did you not program the machine yourselves?"

"Yes... with the labor of one thousand physicists, each spending years at the task. But you must understand, the algorithms are complex, esoteric, and far from intuitive... even for me. Many of the discoveries we made can only be called fortuitous, and none of us had the time to think out what they really meant."

"But surely you can give me some pictures or analogies to explain how the thing works!" exclaimed the philosopher.

"Um... I'm afraid not. Only on the level of 'If you want to predict the value of this particular quantity, perform the following algebraic operations... '," the physicist admitted.

"But that's terrible," said the philosopher. "What have you achieved? Before you built this super-super-computer you had the even-greater computer called Nature, which did the same thing, albeit in a slower and less convenient fashion. If you don't understand the workings of the one any better than the other, you have accomplished nothing!"

"No," said the physicist doubtfully, "that can't be right. We programmed it... so we must have understood it. At least I think so... hmm. Maybe you have a point there. Yes, I think you do... "

And with that the physicist ran off to found the discipline of "superphysics," the science of understanding how the program in the super-super-computer really works. And soon he had an enormous grant, a new Research Institute (of which he was made director), and his name on a hundred new papers. And he lived happily ever after.

"Nature resists imitation through a model," wrote Schrödinger (imagining the thought-train of the opposing school), "so one lets go of naïve realism and leans directly on the indubitable proposition that actually (for the physicist) after all is said and done there is only observation, measurement." Indubitable but not the whole story, thinks the naïve realist. One must try to discover the principles on which an adequate theory can be built. (To the end of his life, Einstein claimed that quantum mechanics lacked firmly laid-down principles.)

Although I may have to revise this passage if I receive any nocturnal visits from angry ghosts, I judge Erwin Schrödinger (when not writing about Vedanta; see below) and Albert Einstein (in his later years, after shaking off the influence of Mach) to have been naïve realists. On the evidence of his 1952 papers, David Bohm was at that time. John Bell always belonged to the faith; so too Edward Nelson. Bohr and Heisenberg were not realists. Feynman's views are opaque to me.

The primary fear of Machian positivists, I suspect, is that without freedom to speculate they will miss out on the great discoveries. But realists also have certain fears. Foremost is that they will be led into a cul-de-sac by their philosophical predilections. There are precedents for such fears; consider, for instance, the lamentable history of "ultramundane matter."

After Newton introduced his theory of gravity, it was widely objected that he gave no explanation for the force of attraction between the planets. *Hypotheses non fingo*, "I feign no hypotheses," was Newton's famous reply. But Cartesian mechanists longed for a model in which the gravitational force would arise from some sort of mechanical action. Such an explanation exists; the first to propose it may have been the French-Swiss scientist George Louis Le Sage, in 1782, but it has been rediscovered countless times since the 18th century.

Here is how it works. (If you have encountered the term "graviton," kindly suppress the thought while reading this section. This other mythical corpuscle will be discussed in Chapter 19.) One supposes that space, while apparently empty, actually contains a flux of tiny, very light particles (which Le Sage called "ultramundane corpuscles") whizzing about in all directions. These particles bounce off the planets and all other mundane bodies billions of times a second. However, we usually do not notice the ultramundane particles, because the tiny kicks they impart on a body occur uniformly (statistically speaking), with kicks from one side canceling those from the other. But an effect may be felt if one body partly or totally screens another from the rain of ultramundane corpuscles. It is not hard to show (it is an exercise in the integral calculus) that the screening effect of two perfectly spherical planets a distance D apart generates an attractive force on them proportional to the inverse of the square of D.

One can imagine an 18th- or 19th-century crank, self-styled revolutionary, or Cartesian scholar leaping from his chair with a cry of satisfaction upon noting this fact. True, there are problems with this model. Mass does not enter properly in the force law without an ad hoc rule on how matter deflects corpuscles. The law of attraction of nonspherical bodies is not the same as Newton's. And there is the lag time in transmitting the force: once the planets start to move, the force on each can only change after a time equal to that required for an ultramundane particle to pass between them. These problems can be ameliorated by adjusting parameters (e.g., speeding up the corpuscles) or simply ignored. Were there not many difficulties—such as the notorious three-body problem—with Newton's theory?

Unfortunately, ultramundane matter adds nothing to our understanding of gravity. If one is prepared, for the sake of "understanding," to complicate the picture, one must also undertake to generate at least one new prediction—ideally, of some phenomenon undreamed of in the old philosophy. The postulate of ultramundane matter leads to just such a prediction, as the astute reader has realized. For a planet under continuous bombardment by fast, light particles is in exactly the same predicament as a pollen grain suspended in water. Like the grain, it should execute an additional motion due to statistical fluctuations in the number of bombarding particles. If the planets only exhibited some form of Brownian motion, then ultramundane matter, like oxygen, would have been a useful hypothesis. Instead, like phlogiston, it was a dead end.

Better to be a privately skeptical supporter of the established paradigm then to willfully violate Occam's proscription against multiplying entities, all in the cause of a philosophical prejudice. But how can one abandon the search for understanding? This is the realist's dilemma.

Now we come to a second axis of disagreement between the principals: the "holism versus reductionism" issue. In his many expositions of complementarity, Bohr emphasized the peculiar "wholeness" or "individuality" of quantum phenomena. "The impossibility of a closer analysis" of the effect of a measuring device on a particle, wrote Bohr in response to EPR, is not due to any special aspect of apparatus but to "a feature of individuality completely foreign to classical physics." This individuality results in turn from the "indivisibility of the quantum of action." This mysterious phrase acted like a charmed amulet in Bohr's philosophy to dispel unwanted possibilities. I do not know if it has a discoverable meaning. David Bohm used similar language on occasion.

The naïve realist may not object to these views on first hearing. When Bohr sites "wholeness" as a justification for instrumentalism, the hackles start to rise. But it is the suspicion that for Bohr "wholeness" has become an *explanatory principle* that really causes misgivings.

The primary source of theoretical progress in science has been the search for unification by means of reduction of disparate phenomena to a single set of causes. In physics, this reduction is usually expressed in terms of a few "fundamental forces" and a few "elementary particles." (Unfortunately, "few" currently means three or four in the first case and dozens in the second.) In biology, reductionism traditionally involves explaining the behavior of an organism on the basis of physico-chemical processes. From Maxwell and Darwin to Einstein and Watson/Crick, the search for a reductive treatment has been the source of all great explanatory successes in science.

But occasionally one encounters a tendency to work "from the top down," so to speak, arguing that a higher level of organization or description can provide explanations for events on the lower levels. In everyday speech we use expressions such as "fear starts the adrenaline flowing," implying such an explanation. But endocrinologists hope to reduce this phenomenon to physiology and chemistry, demonstrating that certain signals from the eyes or ears to the brain cause, let us say, an electrical wave to pass through the cerebral cortex, which triggers the release of a brain hormone, and so on. It is uncommon in this century (at least in physiology departments) to deny that this kind of explanation is possible—but it was quite common in the last. In another context, some Marxist political writers seem to regard "history" as a kind of invisible force acting on individuals or groups.

I can understand explanations based on interactions, even if they involve action at a distance. But I cannot understand explanations of atomic phenomena based on a principle akin to nationalism—as if elementary particles simply "know" their citizenship and so act accordingly. Political scientists think it worthwhile to investigate how citizens are persuaded by their government or by popular opinion to go to war or pay their taxes; if it should someday be proven that electrons are conscious, I would still want to know how they get their news.

The doctrine of complementarity is a form of philosophical dualism; reductionism is the antithesis of dualism. Perhaps it is not surprising, therefore, that Bohr and his disciples became embroiled in the never-ending debate over the virtues of reductionism in biology.

Beginning in the 1930s, many physicists, worried that the final word on atoms had been spoken, were attracted to biology as a source of fresh problems. Bohr encouraged this attitude at his institute in Copenhagen. In 1932, he gave a lecture called "Light and Life" in which he drew analogies between the situation in biology and that in physics before quantum mechanics. Sitting in the audience was Max Delbrück, a 26-year-old German physicist destined to play a major role in modern biology, especially in genetics. (Delbrück won the Nobel Prize in Physiology or Medicine in 1969 for his research on the genetics of bacteriophage, viruses that infect bacteria.) Delbrück set out to find something like Bohr's complementarity principle in biology. His lifelong search for a "fundamental paradox" is both fascinating and rich in irony; moreover, since Delbrück came out of Bohr's school and thought he was pursuing Bohr's epistemological program in another branch of science, this episode sheds light on the nature of that program.

Delbrück fled Germany in 1937, coming to Caltech as a Rockefeller Fellow. At the time he was already well known as a physicist, an expert in quantum mechanics no less, who had made significant contributions to biology. Delbrück and two collaborators had made the first steps toward a molecular model of the gene in 1935, making a proposal subsequently called "Delbrück's model." (They envisioned the gene as a stable form of a complex molecule, and mutation as a jump to another stable form at a higher energy level.) In 1940, Delbrück and the Italian bacteriologist Salvador Luria began phage research at the Cold Spring Harbor Laboratory on Long Island Sound. Their goal in studying this simplest of all biological systems was to solve the mystery of replication: how did the phage particle reproduce itself, producing thousands of identical copies in 20 minutes? Delbrück and Luria would soon make bacteriophage as important to genetics as *Drosophila*, the fruit fly, had been earlier in the century.

In 1949, Delbrück gave an invited address to the thousandth meeting of the Connecticut Academy of Science entitled "A Physicist Looks at Biology." This lecture contains perhaps the most astounding advice ever given by a scientist to colleagues in another discipline. After surveying the most important open problems in biology, Delbrück launched into a reprise of Copenhagenism replete with the usual phrases: "uncontrollable exchanges of energy or momentum," "exclusive experimental arrangements," "individuality," and so forth. Then he made a remarkable comparison. Helmholtz, the patron saint of physicists interested in biology (who discussed living systems from the point of view of physics in the 1870s), had assumed that the behavior of the cell would eventually be explained by the motion of molecules acted on by prescribed forces. Delbrück drew an analogy between this belief and the hopes, now dashed forever (he presumed), that a similar mechanical explanation would be found for the atom.

This lead Delbrück to make a modest proposal. "It may turn out that certain features of the living cell, including perhaps replication, stand in a mutually exclusive relationship to the strict application of quantum mechanics," the fundamental law of matter. "We may find features of the cell which are not reducible to atomic physics but whose appearance stands in a complementary relationship to those of atomic physics." New laws unknown to physics, he suggests, may be needed to explain these features.

If Delbrück relieved biologists of the troublesome task of mastering infinite-dimensional Hilbert spaces, he was not entirely circumspect in his additional remarks. He spoke rather scornfully of the biochemist as one who would persuade us that "the cell is a sack full of enzymes acting on substrates converting them through various intermediate stages into cell substance or into waste products." (Indeed, I remember being so persuaded by my high-school biology instructor 20 years later.) "The vista of the biochemist is an infinite horizon," says Delbrück, and yet

> this program of explaining the simple through the complex smacks suspiciously of the program of explaining atoms in terms of complex mechanical models. It looks sane until the paradoxes crop up and come into sharper focus. In biology we are not yet at the point where we are presented with clear paradoxes, and this will not happen until the analysis of living cells has been carried into far greater detail.

I love that phrase "clear paradoxes," referring to something a scientist should be overjoyed to find.

Delbrück was no fool and pre-emptively dealt with a thought passing through the minds of the startled biologists in the audience:

> Perhaps you will think that such speculations and arguments as here presented are very dangerous: They seem to encourage defeatism before it is necessary and to open the door to wild and unreasonable speculations of a vitalistic kind.

"Vitalism" was—perhaps I should say, is—the view that living tissue is animated by a special principle or force, the "vital force," which cannot be reduced to chemical or physical action. In a sense, it is the antithesis of mechanism. It can be traced back to Aristotle (who called it psyche); in the last century it had numerous adherents. The psychologist and philosopher Henri Bergson (1859–1941) is a more recent proponent of a vital force, which he called by the charming name *elan vital*. He thought it played an important role in evolution. (In the cinema, vitalism seems to be the dominant biological conception: mad scientists ever since Frankenstein have been drawing down the "life essence" from the skies, or draining it from the bodies of their victims.) Vitalism—the postulate of a "ghost in the machine"—represents the great dualistic temptation in the history of biology, and Delbrück knew it.

The denouement of this episode is what makes it fascinating. A few years after Delbrück's lecture, a one-time participant in Delbrück's phage group named James Watson left for Cambridge in England. Watson complained of his "inability to think mathematically," knew no quantum mechanics, and had no interest in complementary thought. He wanted to find a simple account of the gene based on chemistry—just the sort of reductionist program Delbrück had reviled in his 1949 address. In Cambridge, Watson, together with Francis Crick and Maurice Wilkins (and assisted by Rosalind Franklin's crucial donation of data from her x-ray crystallography photographs), discovered a model for gene replication based on conventional chemical mechanisms. This was the celebrated Crick-Watson proposal of base-pairing in deoxyribonucleic acid (DNA), the famous "double-helix." The "secret of life," as Watson liked to put it, had been discovered, and it was garden-variety chemistry. No "clear paradoxes," no new physical principles, no dualism between mutually exclusive pictures.

After the Watson-Crick model proved successful, Delbrück lost interest in genetics and moved into other areas of research. While searching for the ultimate paradox, he had missed out on the greatest discovery in biology of the century. But his faith was unshaken. In a letter to Bohr in 1954, he reaffirmed his belief that when biologists had carried their analysis far enough, the inevitable paradox would be found. "This, of course, had been my ulterior motive in biology from the very beginning."

I do not want to read too much into this episode, as it may represent only one scientist's idiosyncrasy. And both Bohr and Delbrück managed, while "waiting for the paradox," to make Nobel-quality discoveries. But what if biologists had actually taken to Delbrück's views? Is it possible that the subsequent 40 years of unparalleled successes in biochemistry and genetics—leading to gene-splicing, bacteria that secrete human hormones, forensic DNA testing, and a thousand other marvels—might have been replaced by a fruitless search for an ultimate paradox?

There is another aspect to this story. Both Watson and Crick reported that they were strongly influenced in their choice of professions by a little book written by another physicist involved with quantum mechanics:

Erwin Schrödinger. In 1943, Schrödinger, then resident in Dublin, gave a series of lectures published a year later under the title *What Is Life?*, subtitled "The Physical Aspects of the Living Cell." Schrödinger disarmingly remarks that he set out in spite of lack of expertise to gratify his "keen longing for unified, all-embracing knowledge." With the acquired self-confidence of a Nobel Prize–winner, he dives right in, lauding Delbrück's molecular model of the gene as the only possible solution—if it failed, one would have to give up any hope of understanding replication. But it was known from studies on radiation-induced mutation that the gene was fantastically small, perhaps no larger than a few thousand atoms. How could it store sufficient information to specify a complex physiological trait? Schrödinger proposes as an analogy the Morse code, in which two symbols ("short" and "long") in groups of 5 could generate 32 words, and in groups of 10 over a 1000. Even a relatively small molecule, if it had a large number of stable rearrangements, could in principle store a vast amount of information. He was, of course, right on the mark.

(After lauding this "realist" performance, I admit that in the last chapter of his book Schrödinger indulged in a bit of mystical speculation, advancing the dubious notion that the biologist's increasing knowledge of the mechanisms of life will one day lead her to conclude "I am God." He reasoned from the contradiction between free will and determinism:

(i) my body functions as a pure mechanism according to the laws of Nature;

(ii) yet I know, by incontrovertible direct experience, that I am directing its motions, of which I foresee the effects...,

(iii) so I am God, QED. (Presumably, this outburst was a result of Schrödinger's lifelong interest in Eastern religions, particularly Vedanta.)

Reviewing this book, the biologist Hermann Müller remarked: "If the collaboration of the physicist in the attack on biological questions finally leads to the conclusion that 'I am God Almighty,' and that the ancient Hindus were on the right track after all, his help should become suspect.")

There is little doubt that Bohr was a vitalist. Biologists at the time certainly thought so. (Müller after visiting Copenhagen in 1933: "I was glad to visit the physicist Bohr, but I found that his ideas in biology were hopelessly vitalistic.") Bohr in the "Light and Life" lecture wrote:

> ... the very existence of life in biology must be considered as an elementary fact, just as in atomic physics the existence of the quantum of action has to be taken as a basic fact that cannot be derived from ordinary mechanical physics. Indeed, the essential non-analyzability of atomic stability in mechanical terms presents a close analogy to *the impossibility of a physical or chemical explanation of the peculiar functions characteristic of life.*

(My emphasis.) Finally consider the following anecdote, which I heard from a physicist in Princeton. In the early 1950s, Bohr attended a lecture there by a biologist on the Crick-Watson model of DNA. Afterwards there was much enthusiasm expressed for the biochemists' achievement in understanding this ancient mystery, but Bohr had only this to say: "Yes—but where is life?"

The final topic of this chapter concerns Bohr's philosophical agenda not derived from physics. The extent to which Bohr based his views on psychological theories popular in his day is not generally known to physicists; when they hear of it they usually react with disbelief. The naïve realist—who is frequently accused of harboring a philosophical prejudice—is well advised to learn a few of these facts for purposes of self-defense.

There seems to be a reference to psychology in every exposition Bohr wrote on complementarity. In the first post-Como paper, Bohr wrote: "The idea of complementarity is suited to characterize the situation, which bears a profound analogy to the general difficulty in the formulation of ideas, inherent in the distinction between subject and object." In a paper of 1929, he remarked that "the necessity of taking recourse to a complementary, or reciprocal, mode of description is perhaps familiar to us from psychological problems." Later in the same paper there appears the following passage:

> In particular, the apparent contrast between the continuous onward flow of associative thinking and the preservation of the unity of personality

> exhibits a suggestive analogy with the relation between the wave description of the motions of material particles... and their indestructible individuality.

Max Jammer made a credible case in 1974 that this peculiar passage is a direct paraphrase of one in *Principles of Psychology* (1890), by the American psychologist William James (the brother of the novelist Henry James).

William James (1842–1910), who in turn taught physiology, psychology, and philosophy at Harvard, was America's most influential thinker at the turn of the century. His Pragmatism and Radical Empiricism are still widely discussed today. In a chapter titled "The Stream of Thought," James compares consciousness to the flight of a bird: a journey alternating between flights and perchings. The "perchings" represent resting places occupied by sense impressions, and the "flights" are thoughts, which form relations between the impressions. James suggests the virtual impossibility of contemplating such a "flight" at the moment it occurs: "As a snowflake crystal caught in the warm hand is no longer a crystal but a drop, so, instead of catching the feeling of relation moving to its term," we arrive instead at a resting place. "The attempt at introspective analysis in these cases is... like seizing a spinning top to catch its motion, or trying to turn up a gas quickly enough to see how the darkness looks."

The similarity to Bohr's language is apparent, but Jammer further suggests that Bohr's coinage of the term "complementarity" may be traced to James. (I have always thought it suggestive that Bohr adopted a neologism, "complementarity," rather than the available term "duality," which he avoided. Perhaps he reasoned that the "duality principle" would be harder to sell to physicists, with their fondness for unifying everything.) James discussed an experiment performed by Pierre Janet, a psychologist and neurologist studying hysterical disease at the same time as Freud. (They were pupils of the same teacher, the physician Charcot.) Janet hypnotized a patient named Lucie, covered her lap with cards, each bearing a number, and then instructed her that she could not see cards whose numbers were a multiple of three. Upon awakening from the trance, Lucie denied that there were any cards labeled 6, 9, etc., in her lap, but her hand, while she was otherwise engaged in conversation, picked up just those cards that bore multiples of three on their faces. James concluded that "in certain persons the total possible consciousness may be split into parts which coexist but ignore each other, and share the objects of knowledge between them. More remarkable still, they are *complementary*." James even refers to these cases as representing "relations of mutual exclusion" a few pages later. Jammer's case is convincing.

To criticize a scientist for drawing inspiration from another discipline would be anti-intellectual and mean-spirited. But a distinction can be made here. When Charles Darwin imported Thomas Malthus's concept of an endless "struggle for existence" from political thought into biology, transforming it into the "survival of the fittest," he was *borrowing a mechanism* and not adopting a new methodology or epistemology. The creative muse visits so rarely that one welcomes any source of inspiration—from art, music, another science, or even one's dreams—if it helps to uncover *an unknown mechanism*. But if James's musings on the paradoxes of the mind contemplating itself did inspire Bohr, it was not to make a new model of atoms. Instead, Bohr tried to convince physicists they should learn to think that way. (Which raises the question of whether James' ideas proved useful to psychologists. The small sample to which I read the quotations above were visibly annoyed by James' paradoxical remarks.)

There is overwhelming evidence that Bohr was pursuing a philosophical agenda dating from long before quantum mechanics and having little to do with physics. One more selection, this time from a memoir of Heisenberg, should clinch the case. Remembering a sailing trip with colleagues, including a chemist and a surgeon, in which Bohr waxed eloquent on his new interpretation of quantum theory, Heisenberg wrote:

> Bohr began by talking of the difficulties of language, of the limitations of all our means of expressing ourselves, which one had to take into account from the very beginning if one wants to practise science, and... how satisfying it was that this limitation had already been expressed in the foundations of atomic theory in a mathematically lucid way. Finally,

> one of the friends remarked drily, "But, Niels, this is not really new, you said exactly the same [thing] ten years ago."

In conclusion: Bohr's philosophy of complementarity is a dualistic doctrine derived from contemplating mentalistic and linguistic paradoxes. Its supporters made unhelpful and occasionally foolish remarks about biology—and, one suspects, about physics as well.

18. Principles

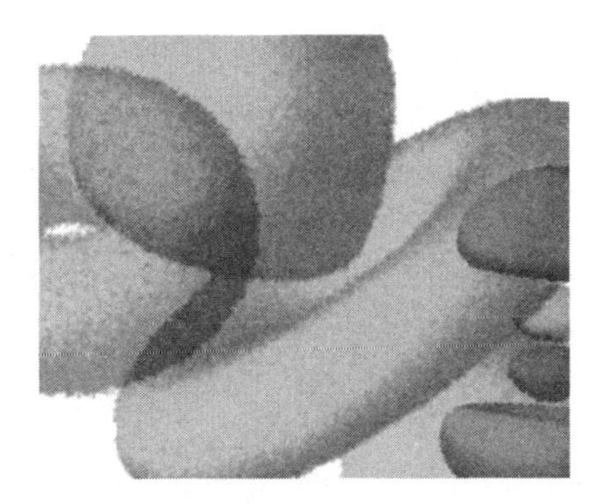

The founding fathers were unable to form a clear picture of things on the remote atomic scale. They became very aware of . . . the need for a 'classical' base from which to intervene on the quantum system. And so the shifty split.

—the late Irish physicist John Bell, speaking in Erice, Sicily, in 1989

What are the principles of quantum mechanics? Since 1925, proponents of various interpretations have championed theirs, while doubters have searched for new ones. Principles, therefore, are important—but what do we mean by a "principle" when talking about a scientific theory?

Some terms, although meaningful or even essential in a discussion, defy exact definition except by instances (e.g., democracy, justice, love, and the reader can supply many more.) So let us begin with a case where the principles were identified with unusual clarity: Einstein's special theory of relativity. In the following I will conflate "principle," "axiom," and "fundamental assumption," since for scientific theories no clear distinctions have consistently been drawn among these terms. Here are the two axioms from which the whole theory follows, as stated in the original article of June 1905 (translated from the German by W. Perrett and G. B. Jeffery):

1. The laws by which the states of physical systems undergo change are not affected, whether these changes of state be referred to one or the other of two systems of co-ordinates in uniform translatory motion.
2. Any light ray moves in the "stationary" system of co-ordinates with the determined velocity c, whether the ray be emitted by a stationary or by a moving body.

The first is the so-called principle of relativity, a confusing name for what is evidently a principle of *invariance*. (Some philosophers at the time became confused by the implied relation with "relativism" as it occurs, for instance, in ethics. There is no connection whatsoever.) The meaning of the principle is simple: no observer in the universe has, by virtue of state of motion, a preferred right to state the laws of physics. The second axiom represents a simple observation about waves that is virtually a part of the *gestalt* evoked by the word. A wave is typically born in a disturbance in a "medium"; but after that the medium takes over and the wave propagates at a velocity characteristic of that medium. So it is for water waves, sound waves, and (by Maxwell's laws) light waves. However, postulating a space-filling "luminiferous aether" to support light waves clearly violates the first axiom. Therefore, in the second axiom, Einstein states what he retains of the wave picture: there should be at least one reference frame in which the velocity of light is independent of the motion of the *source*. (Textbook writers often substitute for the second assumption the ridiculous "axiom" that the light velocity does not depend on the motion of the *observer*, causing the poor student to imagine running along a light wave and... it speeds up? Einstein *derived* this surprising fact from his axioms; if you want to convince other scientists of your theories, put the zingers among the *conclusions*, not among the axioms.)

From these two statements the whole of relativity theory—the slowing of clocks and shrinking of measuring rods in motion, the relativity of simultaneity, $E = Mc^2$, and much more—can be derived. The mathematics need not go beyond high-school algebra, and one even imagines that a very bright student, if told these axioms, would write out the entire theory in time.

As for the general theory of 1915—Einstein's theory of gravity—our "bright student" would first have to learn the non-Euclidean geometry of the 19th-century mathematician Bernhard Riemann, as Einstein did. The principles can then be communicated, but without doing so I present the following historical episode as evidence that they are known.

In July of 1915, Einstein was still having trouble formulating the equations of gravity. At the invitation of David Hilbert, he visited Göttingen to explain the principles he hoped would bring gravity into agreement with relativity. "To my great joy," Einstein wrote to a friend in August, "I succeeded in convincing Hilbert and [Felix] Klein completely." But conveying principles to the greatest mathematicians in the world had its dangers. Hilbert proceeded to derive the correct equations from a variational principle and submitted a paper *five days before* Einstein presented them to the Prussian Academy in November 1915. (Hilbert was so excited he called his paper "The Foundations of Physics" and imagined that he had solved not only gravity but the atomic problem.) Einstein's theory of gravity almost became Hilbert's theory. That winter, Hilbert admitted that he had blocked out Einstein's July lecture, and the two great thinkers patched things up.

As I write, it is seven decades since Heisenberg's epiphany on the beach in the Helgoland, or Schödinger's in the chateau in Arosa, but no one would try the experiment with the "bright student" where quantum mechanics is concerned. Born of desperation in feverish acts of creativity, quantum mechanics is unique in that its equations are known but not its principles—not with anything like the precision one would expect to find in physicists' most fundamental theory. The situation is, as John Bell rightly insisted, a scandal.

Before discussing attempts to find those missing principles, it is worth asking again why Bohr and Heisenberg failed. We have seen in previous chapters that their ideas were ambiguous, ideological (Machian, dualistic, transfixed by paradoxes), and based on circular justifications. And, at least in Heisenberg's case, there was a false analogy drawn with another theory. This is worth a brief digression.

Relativity was that other theory. Heisenberg recalled his motivation for restricting treatment to quantities that can be directly observed as coming from Einstein rather than from Mach (Chapter 3). And indeed there is a Machian component in relativity, as Einstein admitted. In the general relativity papers of 1915–1917 Einstein called "Mach's Principle" the argument that the concave shape formed by the water's surface in a spinning bucket cannot be explained by rotation relative to Newton's unobservable Absolute Space. (The true cause, Mach and Einstein agreed, was rotation relative to the fixed stars.) Special relativity implicitly incorporated several "Mach-like" negative statements such as that time and space are meaningful only when referred to actual measurements with rods and clocks, and that simultaneity between distant places is not a given but must be physically defined. Why then did Einstein quip that "a good joke should not be repeated too often"?

The flaw in the analogy is that quantum mechanics and relativity did not address comparable problems. A theory of the microscopic architecture supporting the macroscopic world will inevitably make reference to unobserved entities—although one can never be sure how long the situation will last. (Reflect on the old saw about never seeing an atom, and Chapter 15. The evidence for quarks inside nucleons, for instance, is entirely indirect at this time.) Relativity treated the relations among macroscopic observables and simply did not face such a challenge. Insisting that these quantities and no others be mentioned in a theory of the microscopic has no precedent in relativity or elsewhere in physics (Mach's philosophy excepted). And, ironically, quantum mechanics did not abolish unobservable quantities; it merely replaced classical ones by others—for example, particle orbits by complex wave functions.

One test that any set of principles must pass (although not alone sufficient to persuade us) is that we can derive the theory's equations from them, rightly understood. The Copenhagen "principles" fail even this test. In expositions, proponents start with Schrödinger's equation or the formula of Born and Jordan or some equally unmotivated notions (see below), and then derive the negative statements about "unknowability," dual pictures, and so forth. To my thinking, it is the reverse of what good scientific reasoning ought to be.

As to the characteristics of good scientific reasoning, there are almost as many views as writers on the topic. Rather than comparing orthodox quantum mechanics to an abstract standard, I turn next to John Bell's critique from an extreme "realist" position, as exposed in his last paper.

"Against 'measurement'" is a broadside against the establishment, the like of which has rarely been seen in the scientific literature. (As any good manifesto should, it set off a debate that has raged ever since in the pages of *Physics World*, the otherwise sedate British trade journal that reprinted Bell's article in 1990.) Bell begins by lamenting the lack, after six decades, of an exact formulation of any serious part of quantum mechanics. By "exact," Bell says, he means not exactly true but "fully formulated in mathematical terms, with nothing left to the discretion of the theoretical physicist." By "serious" he means that it should cover substantial parts of chemistry and physics, and "... that 'apparatus' should not be separated off from the rest of the world into black boxes, as if it were not made from atoms and not ruled by quantum mechanics."

Bell next classifies physicists into those who shrug (the "why bother?-ers") and those who reply: "Why not look it up in a book?" If pressed the former insist:

QUANTUM MECHANICS IS JUST FINE
FOR ALL PRACTICAL PURPOSES

So Bell introduced the acronym FAPP for their characteristic qualifying phrase. These physicists are like those natural geometers, Bell remarked, who were impatient with Euclid's axioms, saying "*of course* in a plane, through a given point, you can draw only one straight line parallel to a given straight line, at least FAPP." (Bell assumed the reader knew the denial of Euclid's fifth axiom was the starting point for the inventors of non-Euclidean geometries in the 19th century.) But is there not something to be said for Euclid?, complains Bell. "... Is it not good to know what follows from what, even if it is not really necessary FAPP?" Bell continues,

> Suppose that when formulation beyond FAPP is attempted, we find an unmovable finger obstinately pointing outside the subject, to the mind of the observer, to the Hindu scriptures, to God, or even only Gravitation? Would that not be very, very interesting?

Having addressed these remarks to the first group, Bell proceeded to the second: those physicists who believe the Measurement Problem—the Gordian knot of quantum mechanics—has been solved within conventional theory. He takes three works (each introduced graciously with a phrase like "indeed a good book") and describes the contortions the authors go through to convince the reader and themselves that quantum mechanics explains measurements. In every attempt the "shifty split" between quantum system and "apparatus," or between quantum system-plus-apparatus and "environment," defeats them. "It is like a snake trying to swallow itself by the tail," remarks Bell. "... It becomes embarrassing for the spectators before it becomes uncomfortable for the snake."

There are many other amusing things in this article, but in my opinion the best is the following:

> Here are some words which, however legitimate and necessary in application, have no place in a *formulation* with any pretension to physical precision: *system, apparatus, environment, microscopic, macroscopic, reversible, irreversible, observable, information, measurement*.

Having employed many of these terms in this book and intending to use more below, I have a vested interest in understanding what Bell meant by this passage. He certainly did not mean that one should ignore the role of the apparatus—quite the reverse. Bell thought the founders, through their emphasis on measurement outcomes—"It would seem the theory is exclusively about 'results of measurement,' and has nothing to say about anything else"—split the world into two realms without having any intention of explaining the interaction between them. Thus "on this list of bad words... the worst of all is measurement." But others offend as well: "microscopic" and "macroscopic" cannot be precisely defined; "observable" should not appear in the *formulation* of a fundamental theory (because, as Einstein emphasized, it is that very theory that determines what one observes); and as for information—"Whose information?" asks Bell.

Although Bell wisely avoided the hot-button "real" in favor of the chilly "precise," his subtext is clear. If we eschew all subjectivist or positivist buzzwords, what is left but representational or naïve realism? What could satisfy us but a realistic account of what is going on in atoms, from which the observables can be *derived*, the boundary between macroscopic and microscopic *uncovered*, and so on?

Bell concluded his article by advertising Bohm's model and another he thought had the best chances of providing that "precise formulation." But before describing these efforts and several others, I detour briefly to discuss certain popular theoretical programs, each blaming our quantum troubles on something other than a lack of physical imagination.

Quantum Logic. A program begun by von Neumann and the American mathematician Garrett Birkhoff in the 1930s, which asserts that, since conventional logic evidently fails us where atoms are concerned, we should have another type to deal with them. Birkhoff and von Neumann characterized the "lattice of propositions" of quantum theory and how it differs from that of classical theories, an interesting result.

Quantum Probability. Introduced by the Italian physicist Luigi Accardi in the 1970s, it supposes that the rules of probability going back to Fermat, Pascal, Bernouilli, and Laplace are wrong where atoms are concerned, and must be replaced by new rules.

The Algebraic Approach. Here I refer to various programs introduced by mathematicians over the years. The claim is that a certain algebra (containing much more complicated objects and rules than you studied in high school) is the proper one to describe Nature; which one depends on the proposer.

Many Worlds. Advanced by the American physicist Hugh Everett III (a student of John Wheeler at Princeton) in 1957, it proposes that anytime anything happens that could have happened differently, the wave function of the universe splits into different "branches." All possibilities are therefore actualities existing "side-by-side," in some sense.

There are other theories in this vein, but these suffice to give the general idea.

"Quantum logic" violates the tenet that criteria for what are good and what are bad arguments should be universal and independent of the transient successes or failures of scientists' theories. (Lately this program's fans seem to have abandoned the notion of alternative logics, so it has become just another way to describe conventional theory.) This moral imperative applies also to "quantum probability," for, as Laplace observed, probability is simply logical reasoning in the presence of uncertainty. Besides, both proposals arise by systematically undervaluing the role of the apparatus. "When one forgets the role of the apparatus," wrote Bell, ". . . one despairs of ordinary logic—hence 'quantum logic'."

The "algebraic approach" can only be loved by a mathematician. Although I am one, I am not willing to elevate mathematical technology above physical principle. On the other hand, the search for a mathematical structure best suited to describing Nature has been a source of progress since the time of the Greek geometers, so I do not entirely dismiss it.

"Many worlds" go, in my opinion, beyond the reasonable and into science fiction. Did I just type "faction" in some universe, and all possible typos in yet others? What can it matter, if I cannot interact with this plethora of worlds? Contemplating "all possible worlds" suggests hubris but is philosophically respectable; claiming that they all coexist is not.

The reader may be irritated by my abrupt dismissals of these popular theories, but as I am profoundly unsympathetic to their goals I feel no compunction to go into greater detail. Interested readers may consult the references in the Notes.

These programs (except perhaps the algebraic one) treat the quantum paradigm with utmost seriousness, as implying a revision of the very rules by which we reason. I suggest reflecting on the historical impermanence of ideas and awaiting the next revolution.

Now I consider schemes that reproduce conventional quantum mechanics, deriving both equations and physical predictions from a list of assumptions. The first is probably the simplest. It is based on a cogent remark by Timothy Wallstrom (then at Princeton, now at the Theoretical Division of Los Alamos National Laboratory) about the heretical theories of Bohm and Fényes/Nelson. It will require a bit of topology theory, but nothing arduous. (Topology is a branch of mathematics that deals with properties of objects that remain unchanged under continuous deformations.)

As we learned in Chapter 10, these particle theories are based on two mathematical quantities: a positive scalar field and a vector field. The scalar field gives the probability for finding the particle in a neighborhood of each point in space. In Bohm's model, the vector field simply gives the velocity of the particle as it passes that point; in the Fényes/Nelson stochastic mechanics it describes something more subtle called the "current velocity," which I cannot go into here. Wallstrom noted that these variables, although computable from Schrödinger's wave if it is known, are not equivalent to having it. There is an interesting additional condition that must be satisfied. But I first need to expound a notion from topology called "simply connected."

Consider a closed curve in some region of the plane. It may be possible to continuously deform the curve to a point, perhaps by shrinking it down around an interior point. If this observation applies to every such curve in the region, one says the region is "simply connected." For instance, the infinite plane is simply connected. But consider the plane minus one point P. If a closed curve surrounds P, it cannot be shrunk to a point *in the region equal to the plane minus P*. The punctured plane is therefore not simply connected. Another example is the surface of a torus, or doughnut.

Wallstrom observed that the velocity field in the alternative theories is not defined where the scalar probability field is zero—that is, where the particle will never be found. And this probability can be zero; for example, in any stable state of the hydrogen atom, the probability is zero at the position of the proton, which partly explains why the electron does not fall into the nucleus despite their opposite charges. The problem comes if cutting out this zero-set leaves space non-simply-connected. This too occurs in the hydrogen atom: there is a stable state for which the zero-set is an infinite straight line. Wallstrom showed that if this happens, the complex Schrödinger wave function cannot be uniquely recovered from the two physical fields. Worse, the equations in the physical variables acquire too many solutions—there is even a continuous family of stable states. This is unfortunate; quantum mechanics was invented precisely to explain discrete stable states of atoms.

It is enlightening to ask what condition restores the equivalence of these theories with quantum mechanics. (Wallstrom's objection applies less to Bohm's theory than to Fényes'/Nelson's, since Bohm did not attempt an independent derivation of quantum mechanics from new principles.) The condition turns out to require an integral of velocity-times-mass (or momentum) around a closed curve to be a multiple of Planck's constant—essentially a continuum version of Bohr's original orbit condition. By starting with the probability and velocity fields—which are natural in a classical particle model incorporating randomness—and adding Wallstrom's condition, one recovers Schrödinger's wave function and with it quantum mechanics. Thus particles plus quantized angular momentum equals quantum mechanics. This is certainly a more intuitive approach than simply starting with infinity-by-infinity matrices or complex wave functions and adding an ideology, but unfortunately the quantization condition is exactly as ad hoc as in the Bohr atom. One yearns for an explanation from new principles.

A second approach combines geometry with statistics. A key observation made in 1980 by William Wootters of Williams College in Massachusetts tells us how to combine the two.

The starting point is the assertion that the geometry of space and time described by Einstein and Minkowski is not the geometry of the microscopic world. Somehow the macroscopic space-time geometry "floats" above a more fundamental geometry, which we deduce from our observations of atoms. The appropriate geometry was characterized by the French mathematician Elie Cartan in 1931 and is called "complex projective space." (Although quantum mechanics was five years old when Cartan wrote his book, he seemed not to know that he was investigating its geometry. Cartan mentions Einstein, with whom he corresponded extensively, but not Heisenberg or Schrödinger.) Let us avoid the abstruse group-theoretic methods of Cartan, however, for old-fashioned analytic geometry; that is, we seek a list of axioms about "points," "lines," and so forth that describes quantum geometry, just as Euclid's described plane geometry.

I have made such a list of axioms—about nine in all, depending on whether some statements are combined. (It is certainly not the shortest possible list.) The "points" of this geometry are quantum states of an atom. The first and most curious assumption is the following replacement for Euclid's first axiom (two points determine a unique line):

Axiom 1

In quantum geometry two points determine a unique sphere.

Given such a sphere, one can choose a great circle defining an "equator" and a 0° "line of longitude." All the states forming this sphere are then described by a latitude and a longitude, or alternatively, by two angles. (One can think of the longitude as a "phase angle" since it runs over 360°; it is here that complex numbers can be brought in if desired, using the argument of a complex number to express the phase.) Indeed, the angles defining this sphere are readily observed in experiments with electrons, photons, magnetized atoms, and other particles.

There are eight additional axioms describing how the various spheres fit together in the ambient space, including one about a "triangle" whose sides are spheres (!), a substitute for Pythagoras' distance formula, and so forth. The others might be postulated on grounds of mathematical parsimony. (For atomic magnetism, i.e., "spin," only finitely many dimensions will be needed, but to include the translational degrees of freedom, one will have to let the number of dimensions tend to infinity.) The question then becomes: where do the statistics come in?

This requires another axiom. Call an observable "simple" if when measured it takes only one of two values, say 0 or 1 (or up or down, yes or no...), with certain probabilities. We introduce the following additional axiom:

Axiom 10

For a simple observable the probabilities are such as to give maximal information about the location of states on the geodesic ("great circle") connecting any two antipodal states.

This assumption using "information," one of Bell's forbidden words, is due to Wootters. I digress to discuss his interesting observation.

Consider the following simple experiment using light, which itself contains an element of mystery. Light passes through a polarizing filter and is polarized in a certain direction. A second polarizer is oriented at an angle with respect to the first. How much light passes through the combined polarizing sandwich? If the angle is zero, the answer is 100 percent: ignoring residual absorption and reflection, there is complete transmission. If the angle is 90°, the answer is zero: the light is completely extinguished. What about the intermediate angles? The light wavc's state of polarization after it has passed through the first filter can be represented by a vector of unit length, which plays the role of the wave function for this problem. To compute the intensity of the light after it passes through the second filter, we simply take the perpendicular projection of this vector along the direction specified by the second filter and square its length. This is the "projection postulate" from Chapter 16 in action. (From trigonometry the answer is: intensity $= \cos^2(\theta)$, where θ is the angle between the filters.) Thinking of light as a rain of little marbles called photons, this result is almost inexplicable.

Wootters found another way to derive this recipe, a way having nothing to do with vectors or wave functions but with *the theory of information transmission*. Suppose one employs a photon counter, an electronic device that detects and counts photons one by one, rather than a photographic plate (or one's own eyes) to measure the light transmitted. The data from a long series of experiments then become a string (Y's and N's) representing passage or absorption of the photon by the second filter:

$$Y\ Y\ N\ Y\ N\ N\ Y\ Y\ N\ Y\ \ldots.$$

Now suppose that we do not know the angle of the first polarizer. (One cannot use ordinary polarizing filters in this experiment because there would be no way of knowing if the photon made it through the first filter. In practice, one replaces the second filter by a Nicholl prism, which reflects or transmits the photon depending on its state of polarization. Both outcomes then refer to the detection of a photon, but at different counters.) Wootters asks us to regard the string of Y's and N's as representing a *message* that Nature is trying to send about the unknown quantity. Being a real number, this parameter represents, in a sense, an infinite amount of information. (For instance, one can identify a real number with its infinite sequence of decimals.) But if nature is restricted to providing information in discrete yes-or-no chunks, the amount we can obtain in a finite number of experiments is limited. Wootters asked: given this restriction, does Nature make the best possible use of her resources?

Wootters proved that the answer to his question is *yes*. Assuming that nature is prohibited from sending messages more cleverly, say by exploiting the order of the symbols to transmit in Morse code, the only information available is the apparent probability of Y, computed from the ratio of the number of Y's to the number of trials. Knowing the formula of quantum mechanics (probability of passage $= \cos^2(\theta)$), we could use a pocket calculator to compute the unknown angle. But this is not the whole story. The communication channel is *noisy*. Suppose we run 100 trials. If we repeat the whole series, our estimate of the probability of "yes" will differ, typically by around one divided by the square root of the number of runs, or 10 percent. (This is the famous $\sqrt{N}$ law for the statistical spread of an average of N measurements, well-known to statisticians.) Our estimate of the angle will differ by a corresponding amount. Wootters made the interesting discovery that the error is smallest provided *Nature relates probability to angle as in quantum mechanics*.

The formula for the correlation in the EPRB experiment likewise gives the most information about the angle between the Stern-Gerlach magnets from the data at both ends. The conclusion is that Nature may be constrained, but she is kind. Combining quantum geometry with information theory gives another set of axioms for quantum theory. Unfortunately, Bell's objection still rings in our ears: "Information? *Whose* information?" And why spheres?

Now we come to approaches that might have pleased Schrödinger, as they avoid both particles and "quantum jumps" entirely. They are "pure wave" theories. Before describing them, however, I address a point about Schrödinger's theory that many people find vexing: what is the square root of minus one doing in a fundamental physical equation?

Answer: it is a red herring. Schrödinger's equation is classical—a fact that surprises many people, including physicists. Every equation involving complex numbers is really two equations involving real numbers; finding these two equations is called "separating real from imaginary." (For instance, the equation $x + \sqrt{-1}\,y = 3 + 7\sqrt{-1}$ just means that $x = 3$ and $y = 7$.) Making this separation, one discovers that Schrödinger's equation is old-fashioned classical mechanics, albeit in very many dimensions, and moreover in the simplest case: combinations of circular motions. (For this observation, which makes quantum mechanics seem almost Ptolemaic, see the Notes.) The $\sqrt{-1}$ plays its usual role of simplifying the algebra; Schrödinger need not have worried about it.

Having disposed of this old conundrum, we come to the sticking point in any pure wave theory: wave-packet collapse after a measurement. Suppose quantum systems are completely described by waves (or quantum states or points in quantum geometry); is there anything preventing us from making an explicit model of an apparatus—also described by quantum states—interacting with and yielding the collapse for the microsystem? This thought occurred to David Bohm and a colleague, Jeffery Bub, in 1966. (It occurred to the author in 1989, who then re-discovered the Bohm-Bub equations. In this field it is hard to have a new idea.) In fact there is no significant obstacle—provided one is willing to go beyond conventional quantum mechanics.

One has to transcend ordinary quantum theory because it is incapable of resolving a dichotomy. Quantum evolution is (quasi-) periodic in time, so it can resolve a dichotomy for at most a finite time interval. Schrödinger's cat *might* become briefly alive or dead in conventional quantum mechanics—but it would inevitably return to its zombie-like state. However, dynamics of another kind, called nonlinear dissipative dynamics, can do the job. The key idea is *multiple basins of attraction.*

Dissipative dynamics can have "basins," or subdomains, of configurations, which all tend ultimately to a fixed, or stable, equilibrium point—like a child's swing settling down to its lowest point after the child departs. Unlike the swing, however, quantum measurements will require dynamics with many such basins. The geometry of the space of quantum states is peculiarly suitable to constructing such dynamics, because it is highly symmetric and has subsets consisting of N points all equally and maximally distant from each other. (See the Appendix.) These points can serve naturally as endpoints for quantum measurements; perhaps this is why the quantum equations seem to be almost shouting the Born interpretation.

In the Bohm-Bub model a measuring apparatus comes equipped with states chosen randomly from all equivalent ones by a "principle of indifference." The interaction with the apparatus puts dissipative forces on the microsystem, N basins of attraction develop, and the microsystem eventually "collapses" (which just means asymptotically approaches) one of N outcome states. The analogy here is pinball—an early and popular application of "chaos." In pinball a small initial randomness is transformed into a discrete variable—the player's score—at the end of the process; the same kind of thing happens in the Bohm-Bub model. Note that in this model the "watched-pot-never-boils" paradox of Chapter 16, "Paradoxes," is resolved, since the measurement clearly will take a finite, minimum time.

After completing their construction, Bohm and Bub had a go again at the hapless von Neumann "impossibility theorem," which, if it ruled out anything, should have ruled out this classical construction. (Recall from Chapter 8, how von Neumann prohibited "hidden variables" in the apparatus by *fiat*.) I do not believe Bohm and Bub were aware, however, of Bell's paper from two years before, which showed that in the EPRB situation either the apparatus acts nonlocally or the two particles act as one.

Another "collapse" model was proposed in 1986 by G. C. Ghiradi and A. Rimini of Trieste and T. Weber of Pavia, Italy. Their scheme was the other one praised by Bell in "Against 'measurement'."

Schrödinger asked in 1935: how does "cat alive *and* dead" become transformed into "cat alive *or* dead" in his diabolical *Gedanken* experiment? GRW proposed that nature includes random "collapses" as well as Schrödinger's continuous law of evolution. This works as follows. Suppose the apparatus is made of N particles. The collapses occur at rate N/τ, where τ is the average time between collapses for an isolated microsystem. The collapse reduces the wave function of apparatus-plus-microsystem to a form without "entanglements." A typical "entanglement" might be between "microsystem has energy one unit and apparatus pointer goes to left on the scale" with "... two... right...," and the collapse takes it to a state representing one or the other possibility.

Now if $N \approx 10^{20}$ and $\tau \approx 10^{15}$ sec (around 30 million years), a collapse will occur to the apparatus (and consequently to the microsystem interacting with it) in $10^{15}/10^{20} = 10^{-5}$ sec. As Bell remarked, in the GRW scheme, "pointers very rapidly point, and cats are very quickly killed *or* spared." An isolated atom, however, will not experience a collapse for a million years or so, so Schrödinger's solution of the hydrogen atom problem is safe.

GRW and others have extended their ideas, and even predicted an additional kind of "Brownian motion" in small grains caused by the quantum collapses going on continually, which might be observable. The primary criticism of their program is that the model introduces several new parameters without a theoretical explanation, and so is a trifle ad hoc. But it is precise, and makes a new prediction.

The big question, as Bell remarked in concluding "Against 'measurement'," is whether any of these precise schemes for quantum mechanics can be brought into agreement with that other great edifice of 20th-century physics: the theory of relativity.

19. Opinions

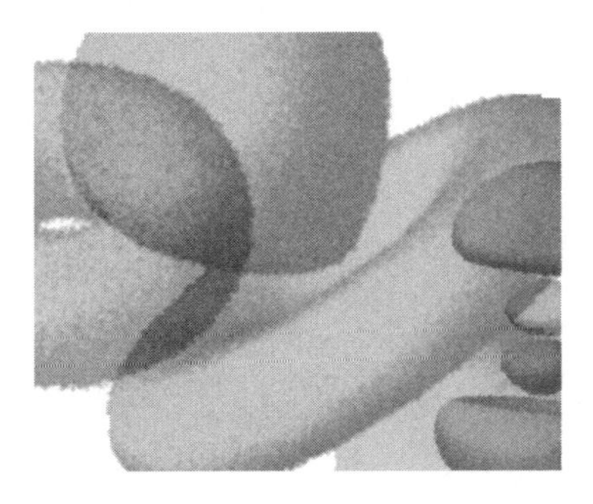

There is no quantum world.

–Niels Bohr to A. Peterson, 1960 (?)

In a theory of the whole thing there can be no fundamental division into observer and observed.

–J. B. Hartle and M. Gell-Mann, 1990

A parable: A man went to the doctor with a fever and a cough. The doctor listened to his chest and then offered a diagnosis: "You have a touch of pneumonia."

"Is it viral or bacterial, Doctor?" asked the worried man, who knew that bacterial pneumonia is easily treatable, whereas the viral type frequently is not. "We will have to take a blood sample and perform several tests," the doctor replied. After phlebotomizing the patient, the doctor permitted him to go home pending the outcome of the tests. The next day the patient returned to hear the following surprising report.

"I'll call the two tests 'A' and 'B' because we don't need to know the details," the doctor began. "We repeated each test twice, so the chance of false positives or negatives is very small. Here are the results: test 'A' showed that you have bacterial pneumonia. Test 'B' revealed that you have viral pneumonia."

The patient knotted his brows in puzzlement over these remarks. "But which is it, Doctor?"

The doctor shook her head. "I'm afraid that is not a good question. We doctors simply accept that sometimes our tests give what we call 'complementary' answers."

"Could I have both infections?" asked the man, not unreasonably.

"No, that has been ruled out."

"But Doctor, I do not understand!" exclaimed the man. "What is my real condition?"

The doctor considered her reply. Finally, she said, "There is no 'real condition'."

For seven decades orthodox physicists have been making claims similar to the doctor's in this story. How you reacted to it likely reflected your own position on the Bohr-Einstein axis. (Positivists, I presume, were not distressed by the doctor's views; realists wanted a second opinion.)

Quantum theorists, in their seminars and articles, tend to state their ideology aggressively, yet one cannot help noticing a worried undercurrent. Consider for example Richard Feynman's repeated use of the two-slit experiment as a talisman against... what? "... a phenomenon which is impossible, *absolutely* impossible, to explain in any classical way," he claimed (incorrectly) in the Red Books. In his popular lectures titled "The Character of Physical Law," given in 1965, Feynman devoted 19 pages to this one experiment, prefaced by the famous mixed metaphor about going down the drain into that blind alley. In substance and omissions, it is a remarkable effort.

Feynman first builds up the experiment as containing the sole mystery of quantum theory: "Any other experiment in quantum mechanics, it turns out, can always be explained by saying 'You remember the case of the experiment with the two holes? It's the same thing'." Then he introduces twin "analogies" of bullets or water waves passing through the holes. He repeats his 1951 claim that probability cannot explain the patterns. There follow sentences about how you might imagine the particle going out one hole and back through the other, or splitting in two, but no one has gotten such a theory to work. Unnamed philosophers are castigated for harboring the prejudice that similar causes should produce similar effects, while physicists are lauded for refusing to speak of certain things. Near the end there are these lines:

> The question now is, how does it really work? What machinery is actually producing this thing? Nobody knows any machinery. Nobody can give you a deeper explanation of this phenomena then I have given; that is, a description of it.

Unable to invent that better explanation on the spot, and battered by the paradoxes, the exhausted reader concedes the game.

Let us take Feynman's statements in this passage seriously. If anyone can find a better account of the two-slit experiment, describe a kind of hidden machinery, then that person has solved the mystery of quantum mechanics. Very well, what is wrong with the pilot wave? Does it not provide a realistic, even mundane description of what is going on? (Recall John Clauser's surfers on the beach from Chapter 13.) But one searches Feynman's 19 pages in vain for the pilot wave, or for mention of Louis de Broglie or David Bohm. Nowhere does Feynman address the principle heresy confronting quantum orthodoxy in this century and say why it was wrong.

Feynman might have said that de Broglie and Bohm did not generalize their theory to include particle creation or annihilation, or to be compatible with relativity. Or that the "pilot wave" for many particles lives in a multidimensional space and so has suspect reality. Making these criticisms would have required Feynman to abandon his central thesis that the two-slit experiment is the only mystery, it is true, although that would have made his remarks even more interesting. Instead, he chose not to speak of his rivals' work. But even popes occasionally made counterarguments to heretics, and one should expect no less from a scientist, especially when the heretics had the stature of de Broglie or Bohm.

David Bohm was Feynman's contemporary and competitor. Feynman helped unite quantum theory with relativity, computed properties of elementary particles to many decimal places, and won the Nobel Prize. He undeniably exceeded his rival as a quantum theorist. Yet Feynman seemed compelled to claim, in overheated rhetoric on many occasions, that no theory like Bohm's had ever existed. It was the specter of Bohmism, I conjecture, that drove Feynman to these excesses.

John Bell was particularly incensed by the establishment's treatment of Bohm and de Broglie. Referring to a popular book by Max Born written in 1949 that Bell read as a student, Bell wrote

> But why then had Born not told me of this "pilot wave?" If only to point out what was wrong with it? Why did von Neumann not consider it? More extraordinarily, why did people go on producing "impossibility proofs" after 1952, and as recently as 1978? When Pauli, Rosenfeld, and Heisenberg could produce no more devastating criticism of Bohm's version than to brand it "metaphysical" and "ideological"?

Bohm's model is usually dismissed with the epithet "classical," which brings me to another pet peeve. The antithesis of "classical" is "modern," and we are all card-carrying modernists (unless you are an artist and prefer to be post-modern). But the term does not mean simply old-fashioned in modern physics any more than it does in painting. Indeed, some theories that postdated quantum mechanics, such as Wiener's theory of Brownian motion (not to mention the post-war heresies), physicists nevertheless call "classical." Nor can one limit its referent to classical mathematics, since (as I have argued in previous chapters) Schrödinger's mathematics were no less "classical" than Bohm's, and more so than Féynes' or Nelson's. So what does it mean to quantum physicists, and why is it the insult of choice?

It may mean "unparadoxical" or "noncontradictory." I believe orthodox quantum theorists reason, consciously or unconsciously, something like this. The microscopic world exhibits paradoxes or contradictions, and this fact is reflected in the best theory describing it. Other theories lack this key element, so they must be "classical."

Bell made a similar point about "complementarity." For Bohr it did not mean what it might in other sciences or everyday life, Bell remarked. Its essence is not multiple views of a whole (like the varying reports of the six blind men of Hindustan who investigated an elephant), but a contradiction. Bell suggested it be relabeled "contradictoriness" or some such term. It is amusing (although admittedly a cheap shot) to substitute such a term for "complementarity" when reading Bohr. For instance, in 1928, Bohr wrote

> Indeed, in the description of atomic phenomena, the quantum postulate presents us with the task of developing a "complementarity" theory the consistency of which can be judged only by weighing the possibilities of definition and observation.

Making the substitution, the problem of judging the theory's consistency takes on a somewhat different complexion.

Hence my opinion of orthodox quantum mechanics, like Bohr's, comes down to the meaning of words. "Classical" and "complementarity," insult and commendation, are euphemisms; the belief concealed is that Nature has been found in a contradiction. But quantum physicists are not simpletons. In their hearts they know such a claim is philosophically unacceptable and would be rejected in the other sciences. The resulting cognitive dissonance is responsible for the bullying tone one finds so frequently in their writings.

Having stated my views (and the narrative having reached the present period), it is time to discuss some interesting current trends. Besides the new theories proposed by heretics (Chapter 18), and the new plans for carrying out experiments, including the original EPRB experiment (Chapter 14), a new "party line" seems about to supplant the old Copenhagen interpretation among establishment physicists. These are welcome developments, for solving the old riddle of uniting the microscopic with the macroscopic is of more importance to science, and of greater relevance to human concerns, than finding a better account of the first nanosecond after the Big Bang or producing yet another exotic particle.

Before coming to the new party line, I sketch some parts of mainstream theoretical physics which the reader may be wondering about. First to "quantum gravity." From one point of view, this is simply the name theoretical physicists give to their ongoing attempts to find a unified theory stretching from nuclei to neutron stars. But the choice of terminology conceals the familiar agenda of quantum mechanics asserting its hegemony over everything. To judge its current status requires a second diversion into the history of relativistic quantum theories, which will also be only schematic.

The first successful unification of relativity with quantum mechanics was quantum electrodynamics ("QED"), the theory of the photon, electron, and positron begun in the 1930s and extended after the war by Feynman, Schwinger, Dyson, and Tomonaga, mentioned in Chapter 10. (Since that time theorists have invented models incorporating two additional forces: the so-called weak force responsible for certain types of radioactive decay, and the strong force that holds nuclei together. These theories led to the successful prediction of new particles but do not seem relevant to the realism problem and so are outside my purview here.) QED relied on "perturbation series," a method of computing quantities by successive approximation. More and more terms are added to approximate the answer, as in the well-known expression for π:

$$\pi = 4 \times \left(1 - \frac{1}{3} + \frac{1}{5} - \frac{1}{7} + \frac{1}{9} - \cdots\right).$$

In the 1930s the individual terms in the QED sums (themselves very difficult to compute) were discovered to be infinite, a distressing situation. Feynman and company succeeded in subtracting off these infinities by a clever algebraic method called "renormalization," a major achievement. That left only the question of whether the series diverges—that is, do the sums eventually oscillate between plus and minus infinity or otherwise misbehave? Although theoretical physicists usually shrug when this issue is raised, mathematical physicists—who are acutely bothered by such divergences—launched a research program called "axiomatic quantum field theory" in the 1950s with the goal of ameliorating the situation. The American mathematical physicist Arthur Wightman of Princeton University proposed axioms these theories should satisfy, axioms that therefore represent another set of principles for quantum theory. (The Wightman axioms postulate Hilbert space without further discussion and so are not an attempt to restore realism as defined here.) This axiomatic program has had major successes, such as the proof of the combined space, time, and charge reversal symmetry of all relativistic quantum theories, a spectacular result. But the rigorous construction of a theory of interacting particles has remained elusive.

For the much more difficult Einstein/Hilbert equations describing the gravitational field, the Feynman/Dyson/Schwinger perturbation techniques simply do not work. (In the 1970s I attended seminars at which students of Feynman or Schwinger—the next generation—aggressively asserted that "gravity is merely a spin-two quantum field theory." When I challenged this remark by asking for a computation of space-time curvature, a phrase like "it will appear at a sufficiently high-enough level of approximation" would be casually tossed out. These simplistic hopes have since evaporated.) Although many approaches have been suggested, at the present time a unification of quantum mechanics with gravity remains an Eldorado, and a "graviton" a mythical beast. Not a single experimental prediction of any "quantum gravity" theory exists today; meanwhile, general relativity survives one experimental test after another. (Recall that in 1915 Einstein had already proposed *three* such tests.)

Proponents of other generalizations of quantum field theory, such as the currently popular "string theories" in which zero-dimensional particles are replaced by one-dimensional "strings," have claimed success in unifying quantum theory with gravity. Because of the technical difficulty of these theories, I cannot venture an opinion. (I have heard, however, that string theories have an *infinite* number of free parameters, which makes me suspicious.) As I write, the string theorists have yet to produce a prediction for any experiment likely to be attempted in the laboratory; nor have they pointed to any observational data, e.g., from astrophysics, that they hope to explain. The theory seems primarily meant to remove some of the scandalous infinities from conventional quantum field theory, a laudable but limited goal.

Despite these setbacks, many theorists are convinced that gravity is the key to resolving quantum paradoxes such as the collapse of the wave packet. Arguments for this position have been made, for instance, by two British scientists, the theoretical physicist Steven Hawking and the mathematician Roger Penrose. Penrose in particular made very serious attempts in the 1970s to find the long-sought theory with his "twistor" program. (See his book *The Emperor's New Mind* (1989). He does not judge his program to have been entirely successful, however.) Gravity is "the only other game in town," so I sincerely hope one of these programs succeeds—and that its success does not rely on our not speaking of certain matters.

Now I come to the new establishment party line, although in these unsettled times "establishment" needs clarification. Unlikely as it may seem, the term can be given a useful definition. An establishment physicist is one who believes a full account of the universe can ultimately be found within the mathematical confines of conventional quantum mechanics—meaning a linear Schrödinger-type evolution equation together with a suitable interpretation. "Heretics" are those who insist we must use different or supplementary mathematics. This definition makes a fairly clean division of the historical players, putting EPR, Bohm, Bell, Nelson, GRW,... on one side and the physicists to be mentioned below on the other, although admittedly leaving Schrödinger sitting on the fence.

The new establishment position is called "decoherence." The idea goes back to von Neumann, who suggested it as part of his treatment of the Measurement Problem. It has been developed particularly by cosmologists, who hope to apply quantum theory to the whole universe. Cosmologists are naturally unable to adopt Bohr's subjectivist views, being constrained by the Copernican principle and the great size and age of the universe not to put human observers at the center. Decoherence might be described as the State Department approach to the problem, trying to resolve conflicts without inventing new institutions. This program has two problems to solve.

The first arises from paradoxes inherent in the quantum description of possibilities by complex amplitudes (synonymous with wave "heights") discussed in Chapter 16. The American mathematical physicist and statistical-mechanics expert Robert Griffiths of Carnegie-Mellon University achieved a better understanding of these paradoxes in 1984. Griffiths asked whether conventional quantum mechanics permits one to talk of "the history" of a quantum system, as we can for a classical system. (Feynman introduced a "sum over histories" interpretation of quantum mechanics in 1949, but, because of severe mathematical difficulties, it is not clear whether it works in every case. Besides, the various histories are all included at once, so the term "history" cannot have its usual meaning.) The answer is: sometimes.

Consider a compound two-slit experiment, as in Figure 28. A particle can pass through two screens by way of slits s_1 or s_2 in screen one, and s_3 or s_4 in screen two. Suppose that the probabilities that the particle passes through s_1 or through s_2 each equals 1/2. If classical mechanics held for the experiment, it would follow that

The probability the particle passes through s_4

is equal to 1/2 times

the probability the particle passes through s_4 *after passing through* s_1,

plus 1/2 times

the probability the particle passes through s_4 *after passing through* s_2.

In quantum mechanics, this may fail due to wave interference, which can make the first mentioned (unconditional) probability zero, while the others are positive. (For example, suppose the complex amplitudes associated with the paths leading to s_4 are: s_1 to s_4, $0.3 - 0.4\sqrt{-1}$; s_2 to s_4, $-0.3+0.4\sqrt{-1}$. To compute the quantum probabilities, one adds the complex numbers associated with the event described and squares the length. Thus, the first mentioned is zero, and the second two sum

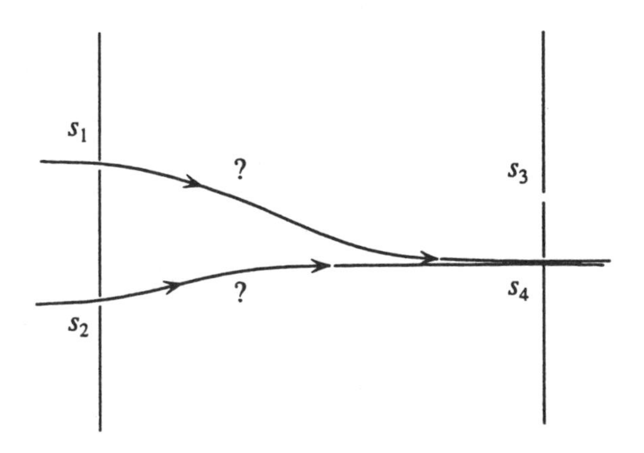

Figure 28

to $(0.3)^2+(0.4)^2 = 0.25$.) Recall how physicists such as Feynman dealt with this conundrum by invoking UPCAM at the slits if the paths are observed. How might this work, exactly? Perhaps some outside agency, such as the measuring device or the "environment," comes in and scrambles the phases of the complex numbers associated with the different possibilities. (Thinking of a complex number as like the hour hand on a clock, this might mean giving the hand a random twist.) The interference due to coherence of the complex waves propagating along the two paths then disappears, and ordinary probability rules hold. Thus "decoherence." Histories that are permissible are said to "decohere," and it must be explained why we see only these in the world around us.

The new establishment proposal for solving this problem violates all of Bell's proscriptions with a vengeance. The supporters of the program—among theoretical physicists in this country proponents include James Hartle of the University of California at Santa Barbara, Murray Gell-Mann (Nobel Prize winner and inventor of quarks) of Caltech, and Wojciech Zurek of the University of Texas, Austin—try to find "environments" that if averaged over result in decohering histories of macroscopic objects. One possibility often mentioned is the "sea" of blackbody photons left over from the Big Bang, which computations show can decohere a dust grain in a nanosecond—explaining why we never see the grain in two places at once. (I asked Professor Hartle after a lecture he gave in Seattle whether these photons might recohere something somewhere else, say a dust grain on Mars, but we failed to connect.) Unfortunately, which variables represent the "environment" and which the "system" is not specified in standard quantum theory. So the theorist has the liberty to choose which parts are to play which role. "For all practical purposes" (FAPP), it may work.

The proponents of this theory have a second dilemma to resolve, however. In addition to Which Part of the Universe Looks Classical?, there is Why This Particular History? Decoherence may explain why Schrödinger's cat has a definite state of health, but which one is it? At this point the Initial Conditions of the Universe are usually invoked to select out the Classical World, with the implied hope that some form of classical mechanics will govern the evolution of the Classical Variables as well.

With its tricky cancellations and cosmic ambitions, success of the decoherence program will demand great ingenuity from theorists. However, as it is still mired in the "shifty split" and leaves many choices to the theorist, John Bell would not have regarded it as the precise account of atoms he longed to see. I also do not believe one should call on the (unknowable) Initial Conditions of the Universe to explain any properties of the world. But it is remarkable how these theorists so readily slough off 70 years of complementary thought. Bohr fought hard for his principles, yet these physicists do not find them any more plausible than do biologists, psychologists, or other scientists. Although I share Bell's outlook and hope for something better, this represents at least a partial return by mainstream physicists to the common scientific fold—in my opinion.

20. Speculations

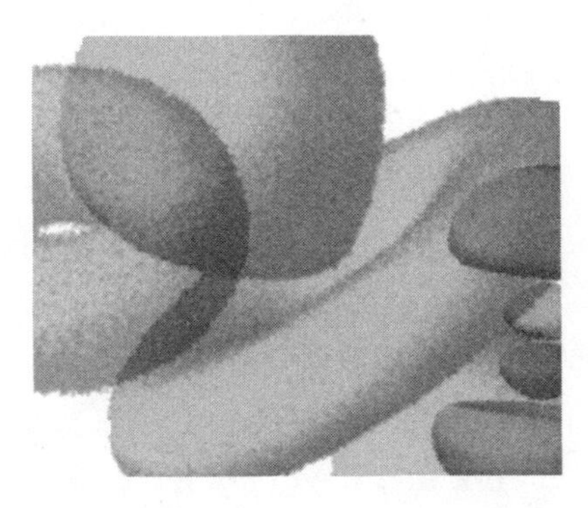

Can we guess the shape of theories yet to come? History dampens our expectations, with quantum mechanics itself throwing the coldest water. But indulging in some harmless speculations can be entertaining and may help clarify some issues. Before the free-associating begins, however, there are some needed preliminaries, which will make this chapter a bit of a hodge-podge. First, a tighter focus on material from previous chapters about determinism and locality will be useful. Then we require a few additional points about quantum mechanics that did not fit conveniently into the historical narrative. Finally, we come to the shameless speculating.

Recall again Einstein's famous aphorism: "God does not throw dice." Einstein may have genuinely disliked giving up determinism, but his principle objection to quantum mechanics lay elsewhere. (Einstein was no piker about exploiting statistical methods. Analyzing fluctuations was one of his favorite tricks.) "That which really exists in [region of space] *B*," Einstein wrote to Max Born in 1948, "... should not depend on what kind of measurement is carried out in part of space *A*; it should also be independent of whether or not any measurement at all is carried out in space *A*." But quantum mechanics, with its "spooky actions at a distance," violated this requirement. "My instinct for physics bristles at this," Einstein told Born. Realism and locality—the central issues in the debates with Bohr—were Einstein's primary concerns.

In 1964, Bell put an entirely new spin (excuse the pun) on the questions Bohr and Einstein debated in 1935. Bell demonstrated that the EPRB experiment could not be explained on the basis of local realism, the postulate that there exist enough "facts" in each region to determine the outcome of the experiment performed there *and any others that might have been performed*. He used EPR's own thought experiment to refute an assumption underpinning their "criterion of reality": knowledge of what might have taken place at *B* on the basis of information obtained at *A* does not permit the deduction that additional "facts" exist at *B if action at a distance connects these two parts of space*. Einstein dismissed this possibility on the basis of instinct, and Bohr concurred. ("There is no question of a mechanical disturbance of the system under investigation during the last critical stage of the measuring procedure," wrote Bohr in reply to EPR.) Battling over realism and positivism, these two great physicists missed the most interesting aspect of EPR's *Gedanken* experiment.

If we are positivists, there is no cause to speculate. So let us be realists but admit some form of nonlocality into our system. Must we accept what Einstein could not—that what happens at *B* may depend on what someone does at *A*? It all hinges now on our faith in determinism. For as we learned in Chapter 12, "Dice Games and Conspiracies," there is a trade-off between giving up determinism and abandoning locality—between believing in a dice-throwing God and admitting spooky actions at a distance.

If one grants the Old One the right to play dice, the distant action that Nature must contain shrinks in significance from particles shaking hands across the light years to a mere statistical regularity in their behavior. Metaphorically, one can think about the situation this way. Our universe is being "selected" from all possible universes on an ongoing basis by a random scheme such as throwing an enormous number of dice. (I am *not* suggesting here that all possible universes exist side by side, as many-worlders assert. Use of the phrase "all possible chess games" does not imply that somewhere 10^{100} chess boards are in play.) This dice-throwing precludes effects from regularly following causes. So any dependence of events at *B* on what someone does at *A* can only come out in the statistics. We set up our EPRB experiment and let a friend choose the apparatus setting on the right, while we watch the particle go up or down on the left. We repeat it again and again. If active locality holds, we see no trace of the friend's actions at *B*. With justification we can claim that what someone does at *A* is irrelevant to what happens at *B*. (Recall from Chapter 12 how carefully one must treat the phrase "what someone does at *A*"; there may be statistical dependence on what *occurs* at *A*, especially events we cannot control, without implying causation.) Einstein might not have objected so strenuously to this scenario.

Can one know with certainty that the Old One is rolling the bones? Unfortunately the answer is no. Mathematicians and others interested in "chaos"—that misnomer for the study of highly unstable deterministic processes—have learned how to construct machines that behave as "randomly" as one likes. If we could study such a machine in complete detail and *in complete isolation* for a long enough time, we might "solve its mechanism" and hence reject dice-throwing. But Bell showed that completely isolating a system from the rest of the universe is probably impossible. Alternatively, we might study two apparatus correlated as in the EPRB experiment. We could then deduce stochasticity, but only by assuming the world adheres strictly to the postulate of active locality. Unfortunately, Nature might be nonlocal on some hidden level.

A philosopher concerned for the principle of sufficient reason will prefer hidden action at a distance to pure dice-throwing. A physicist reared on relativity and quantum mechanics will likely go with the latter. Perhaps in the end it is a matter of taste.

Whether one tries for a stochastic theory or insists on determinism, the chief obstacle to progress seems to be understanding how the space-time continuum, Minkowski's "world-hypothesis," can be compatible with the failure of locality. We sometimes suffer the illusion of space-time separation in circumstances where Nature ignores it. It is tempting to imagine that physicists, in constructing their particle or wave models (both, after all, fundamentally rooted in the space-time continuum), are like game-players at the video arcade. There the players watch and partially control various little creatures, and perhaps walls or barriers constraining them, on a two-dimensional screen. If asked to construct a theory to explain this universe, would not our games-playing physicists naturally construct models of creatures ("particles") or boundaries ("wave fronts") obeying various rules? How surprised our juvenile theorists would then be if the screen were removed to reveal the world behind the appearances—the magnets, cathode-ray emitters, and silicon chips—in which no two-dimensional entities at all can be found (except as coded in a very complicated way in computer memory)? (The "quantum geometry" axioms in Chapter 18 were one attempt to understand what might lie behind the screen.)

Without trying to construct a theory of this concealed machinery, it is useful to exhibit a simple "hidden-variable" model of the EPRB experiment, just to see what is involved. This is not difficult. Recall the picture of the hidden variable adopted in "Bell's Theorem" as a point in the unit square. (I presume the fastidious reader convinced herself or himself that this pictorial representation was inconsequential for the argument.) I mentioned in that chapter the possibility of allowing the circles representing what happened at the left analyzer to move about as one rotates the right analyzer and how this would represent action at a distance. One can build a model of the EPRB experiment in this way, but it is simpler to make *the space of hidden variables itself* depend on the angular deviation between the two analyzers.

Here is how it works. First divide a larger square, say of side two, into four quadrants (see Figure 29) representing the outcomes "up-up," "up-down," "down-up," and "down-down." That is, if the hidden variable happens to lie in the lower-left quadrant of the square, the particle goes "up" on the left and "down" on the right, and similarly for the other quadrants. Now erase part of the square, with the section to be erased depending on the angle between the two analyzers. In Figure 29 show the regions to be erased for angular deviations of 0°, 90°, and 120°. In each experiment, the hidden variable is chosen by dropping a thumbtack, provided it lands in the figure left after the indicated part is removed. Of course, there is some arbitrariness in how to draw the figures.

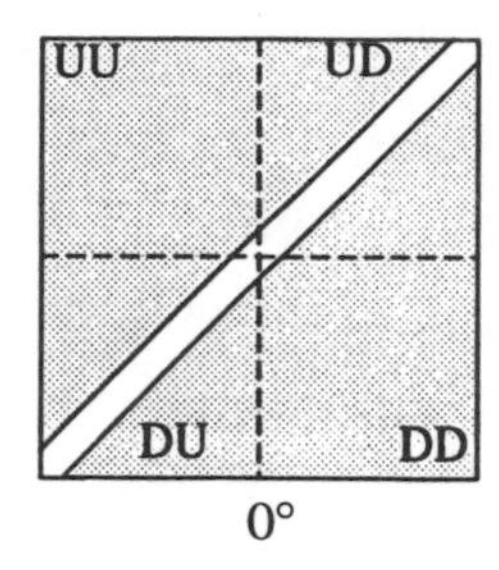

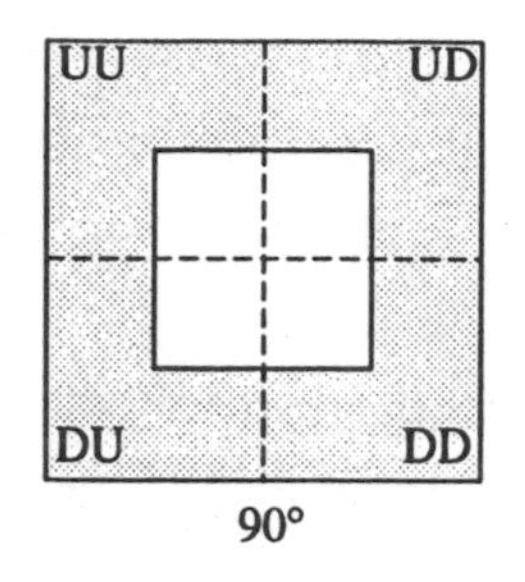

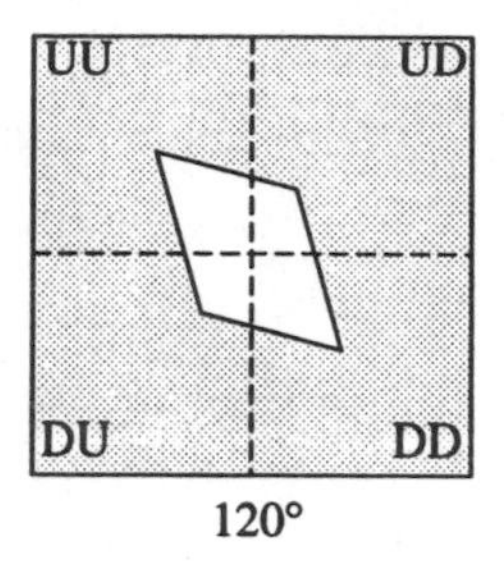

Figure 29

This model makes it easy to see that rotating one analyzer cannot affect the results at the other one, statistically speaking. This is due to a *symmetry property* of all the figures in the drawings. Draw a vertical line bisecting the figure and reflect it as if through a mirror held perpendicular to the page and parallel to this line. Then draw a horizontal bisector and reflect again. After these two operations, the figure will be restored. Since the first reflection interchanges the probabilities of up and down on the left while the second leaves it unaltered, these probabilities must be 1/2–1/2, and similarly for the right detector.

What is the meaning of this construction? The existence of the apparatus, it seems, transforms the hidden variables describing the particle pair. Either these variables are altered deterministically depending on the settings of *both* analyzers (violating active locality on a hidden level), or Nature throws dice in a way depending on both (violating passive locality). It is almost as if there is only one particle present, either existing simultaneously at both places, or jumping instantaneously back and forth, sampling both environments. But to call this a "theory" is to confuse pictures with paradigms, metaphors with mechanisms.

This simple model also helps make it clear why the "no-go" theorems proved in this century (see Chapter 8) left realism and determinism unscathed. These theorems were not entirely void of content; they ruled out a class of models called "noncontextual hidden-variable theories." If the reader can tolerate another doctor story, consider the following (real-life) experiment. Medical researchers divide a group of heart patients into two groups (by a coin flip or by other means). One group receives coronary-bypass surgery to improve blood flow to the heart. The other receives drug therapy. The researchers then follow the patients for a number of years, hoping to learn which group does better and thus which treatment is superior.

Nowadays such a clinical trial is a large cooperative venture among doctors, medical researchers, and statisticians. It is the statistician's job to make a model of the clinical scenario to assist in data analysis. As a matter of course the statistician will treat the factor controlled by the experimenters (here surgery versus drugs) as a "controllable parameter" rather than a "random variable" as discussed in Chapter 12. (The statistician's jargon for these are "covariates" and "outcome variables," respectively.) Provided the two groups show a statistically significant difference, the statistician will enter the parameter into the model, which will then make different predictions for survival or quality of life in each group. This is normal scientific practice, and the statistician would be profoundly puzzled by an insistence that every factor, controllable or "random," be treated in the same way. ("Why ignore the covariates?" she would ask.)

Yet this restriction was unconsciously made in every impossibility theorem proved in this century, except Bell's (who knew the "hidden assumption" in his analysis of EPRB involved locality). The authors of these theorems compared quantum mechanics to an unreasonable standard, a hypothetical "universal probability law" describing everything done or undone, observed or hidden. (See Faris' Appendix and in the Notes section for the technical issue involved.) I suspect that those who interpreted these theorems as ruling out "hidden variables"—code words for a real world—had insufficient practical modeling experience to appreciate this point.

Quantum mechanics has another peculiar feature, which is important to grasp when thinking about alternatives. This is the way systems are combined in quantum theory. But first let us ask, how do we put systems together classically?

In Figure 30 (a)–(f) I have drawn two peculiar "classical" objects, which might be made from colored wires and lumps of clay. We ignore the positions and velocities of these objects; all we are interested in is their "internal states." These are given by the status of four wires—two of solid color and two striped—sprouting from each object. The objects are always found with these wires connected, as in Figure 30. Hence each has two possible states: same to same (call it "SS") or same to different ("SD").

Now consider the two objects together. How many states are possible for the combined system? Clearly, the number is four: (SS, SS), (SS, SD), (SD, SS), and (SD, SD). In mathematical lingo one says the configuration space of the combined system is the Cartesian product of the individual ones. There are four because the number of elements in a Cartesian product of two finite sets always equals the product of the number in each. This is the classical picture (for discrete objects). Now look at (e) and (f) in Figure 30, in which the two lumps have conjugated into new, "nonclassical" configurations. There are two new states of this type, for a total of six. Thus the set of combined states is no longer a Cartesian product of the separate ones.

A similar situation transpires in quantum theory: the space of quantum states of two particles is larger than the Cartesian product of their individual states. The "strange entanglements" that give rise to Bell's theorem derive partly from this fact. (See Faris' discussion in the Appendix.)

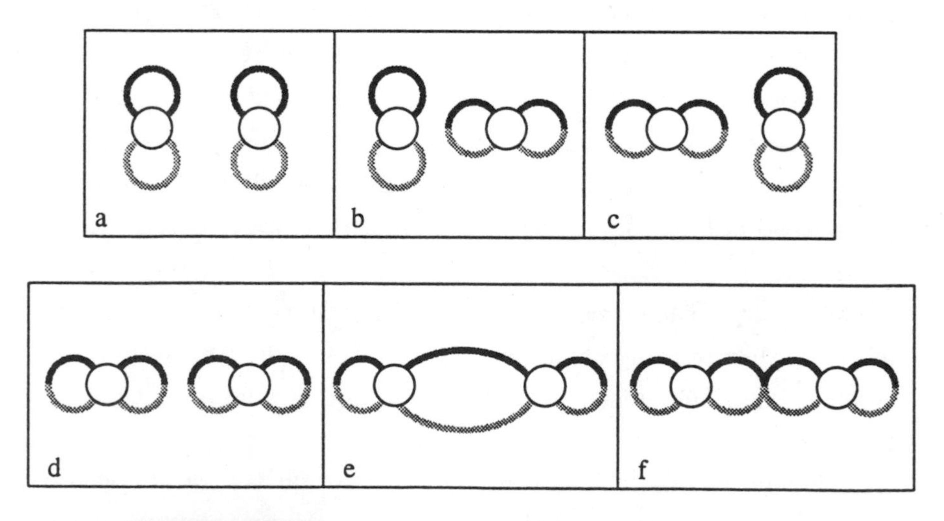

Figure 30 (a)-(f)

These wired lumps of clay might serve as a metaphor for the "embrace" of distant particles.

We are almost prepared, but first: what is our goal? To find a realistic account of elementary particles, at least in some important cases. We reject Bohr's counsel to think in complementary pictures, for realism is the engine that powers our imagination. (Experiments are the rails that keep it on course.) We accept Bohr's insistence that we fully describe the apparatus and Feynman's about how UPCAM acts on a particle observed. But we shall explain the perturbation precisely, with no freedom left to the theorist. (If the reader gasps at the audacity of this program, I can only remark that if no one searches for the Philosopher's Stone, it will never be found.)

The time has come for speculation. The theory that has been missed is completely deterministic, similar to Bohm's. Hilbert's space and Schrödinger's linear equation are part of the picture, but there are also higher-order, nonlinear terms. Particles interact as in the clay-and-wire model, although in a more complicated way. The apparent dice-throwing is due to the long-range interactions of identical particles (their "wires" being infinitely stretchable), which prevent a single particle from being observed in isolation. If an apparatus is present, these interactions are also responsible for the "collapse of the wave function." Active locality fails on a hidden level.

Then again, perhaps the randomness is not due to the particle interacting with every like particle in the universe. It is due to deterministic "chaos" generated in the interaction of the microsystem with the apparatus (as suggested in Chapter 18).

But why insist on determinism? The missing theory is stochastic, like Féynes'/Nelson's or GRW's. Randomness is the Old One throwing dice, and It chooses to violate passive locality.

On the other hand...

Postscript

This century began with a dilemma and a paradox. Two great theoretical paradigms—Newton's mechanics and Maxwell's electromagnetism—each supporting a splendid structure but contradictory at the join, formed the dilemma. The twin pictures of wave and particle cast the paradox. Now as we approach the century's end, despite all the successes, we face, strangely, an almost identical situation. Two successful theories, general relativity and quantum mechanics, are triumphant in their own realms, yet remain strangely silent across their mutual boundary. And our old friend, the paradox of the continuous and the discrete, remains.

Quantum mechanics was one of the most successful creations of the human intellect. From the color of neon lights to the hardness of diamonds to the magnetism of electrons, it correctly described a host of physical phenomena. When it worked, it worked. But quantum mechanics was not the end of physics. I had a great time writing this book, and I hope the reader—whatever his or her philosophy—enjoyed reading it. But I am, of course, dissatisfied. In the last two chapters I advanced many speculations, some no doubt woolly. I have tried to implement some of these speculations mathematically, with indifferent success.

It would give me the greatest pleasure if a reader were to take one of these vague speculations about Bell's "precise theory" and get it to work. No doubt my concepts will prove inadequate, so by all means invent new ones; they need be neither "quantum" nor "classical." State your principles, then use the right mathematics, no matter how complex or abstruse; if you are right, others will be only too happy to learn your methodology. Remember that you will first have to explain why quantum mechanics worked so well—just as Einstein had to explain, in 1915, why Newton's theory was so successful. Be sure to make at least one new prediction for a plausible experiment, preferably one undreamt of in the old philosophy.

You will have a lot of trouble getting published, and even if you do, you will encounter extreme skepticism. So check every detail again and again, and if your theory does not work, discard it. Remember, as Richard Feynman once remarked, the easiest person to fool is yourself.

Good luck.

Appendix

Probability in Quantum Mechanics

by William G. Faris

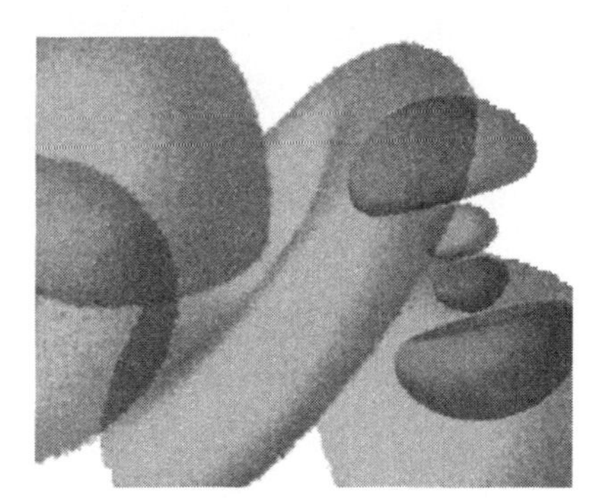

INTRODUCTION

The purpose of this appendix is to give a precise mathematical formulation of Bell's theorems at a relatively elementary level (basic probability, trigonometry, no calculus). This will allow the reader to form his or her own judgment about the implications of these results for our view of Nature.

This enterprise will require a preliminary explanation of probability and of quantum mechanics and of their relation to each other. The probability part presents the elementary theory and then elaborates the notion of passive locality, the requirement that dependence between simultaneous distant events must have an explanation in terms of prior preparation.

The quantum mechanics part begins with a sketch of some of the most important general features of quantum mechanics. This is followed by a description of a spin $1/2$ quantum mechanical system. This system is very special in that it is possible to give a probability model of the system that reproduces the results of quantum mechanics.

The quantum mechanical system of particular interest is a system of two spin $1/2$ particles in a state of total spin zero, the singlet state. Bell's first theorem shows that there can be no probability model for this system. It might be, however, that there were several probability models corresponding to various choices of what to measure. Bell's second theorem shows that, under the assumptions of active locality (no instantaneous influence over long distance) and of passive locality, this is also ruled out.

PROBABILITY

Probability theory comes in if we are unsure of what will happen in an experiment. This could be due to our ignorance of some facts, or perhaps God really is a dice player. However, even though repeated experiments do not always give the same result, at least there is a well-defined notion of the outcome of an experiment.

The fundamental concepts of probability theory are outcome, event, and probability. An event is a set of outcomes, and the probability of an event is intended to give a theoretical prediction of the proportion of times that the outcome belongs to the event, when the experiment is repeated many times. There is an established mathematical framework for probability that gives these notions a precise definition.

Outcomes In this framework there is a set Ω of possible *outcomes* of the experiment. We think of the experiment as something that could in principle be repeated. Each time the experiment is repeated, one of the outcomes is the result.

Example: The set of outcomes of an experiment of tossing a pair of coins may be represented by the set $\Omega = \{HH, HT, TH, TT\}$. Each time the experiment is repeated, exactly one of these outcomes occurs.

Example: The set of outcomes of an experiment of firing at and hitting a circular target with radius one meter is represented by the set D

of all points in a plane of distance not exceeding one from a fixed point. Each time one shoots at the target, the outcome is a particular point in the disk D.

This notion of describing the result of an experiment by a single set of possible outcomes is the most elementary idea of probability theory. It is precisely this notion that is violated in quantum mechanics.

Events Certain subsets of Ω are called *events*. Each event E corresponds to a possible question about the result of the experiment; the question is whether the outcome belongs to the subset E. The empty set $\varnothing$ is called the *impossible event*, since no outcome is an element of it. The set Ω itself is called the *sure event*, since every outcome is an element of it.

Events may be combined to form new events by logical operations. If A is an event, then the *complementary event* **not** A consists of all the outcomes that are not in A. If A and B are events, the *union* event A **or** B consists of all outcomes that are in A or are in B. The *intersection* event A **and** B consists of all outcomes that are in A and are in B. Events A and B are *exclusive* if A **and** B is impossible.

Often one wants to consider a collection of events that may be freely combined by these logical operations. We shall use a special terminology for this. A collection of events is a *closed system* if whenever events are combined by the set operations of complement, union, and intersection, the resulting sets are still events in the collection. In other words, the collection is closed under these logical operations; they do not lead out of the collection. (In advanced probability a closed system of events is often called an "algebra" of events or a "field" of events.)

If a closed system of events has only finitely many events, then it is determined by a *partition* of the set Ω. This is a collection of nonempty exclusive subsets of Ω whose union is Ω. The other events in the closed system are obtained by taking unions of these events. We say that the closed system is *generated* by the events in the partition. If there are k events in the partition that generate the closed system, then the closed system itself has 2^k-events.

Consider, for example, the case when $k = 3$. Suppose the three exclusive events in the partition that generate the closed system are A, B, and C. The closed system generated by A, B, and C consists of the impossible event, the original events A, B, C, the three union events A **or** B, A **or** C, B **or** C, and the sure event. Indeed there are $2^3 = 8$ events in the closed system.

A more concrete example may help. The experiment of tossing two coins has four possible outcomes. The set of outcomes is $\Omega = \{HH, HT, TH, TT\}$. The largest possible closed system of events for this experiment is all subsets. This closed system is generated by the partition into 4 events $\{HH\}$, $\{HT\}$, $\{TH\}$, and $\{TT\}$ corresponding to the outcomes, and all 16 subsets of Ω belong to this closed system.

Sometimes one is interested only in partial information about the outcome of the experiment. For instance, we might only be interested in the total number of heads. This information is contained in a smaller closed system of events, in this case generated by a partition into the three events $\{HH\}$, $\{HT, TH\}$, $\{TT\}$. Knowing which of these events happens is equivalent to knowing whether the number of heads is 0, 1, or 2. Other events in this closed system are the event $\{HH, HT, TH\}$ of one or more heads, the event $\{HH, TT\}$ of an even number of heads, and the event $\{HT, TH, TT\}$ of no more than one head. The impossible event and the sure event also belong to the closed system, so there are a total of eight events in the closed system.

This example makes clear that specifying a closed system of events is the same as specifying a consistent set of questions about the result of the experiment. Another example of a closed system of events corresponds to asking questions only about the first toss. (Information about the second toss may not yet be available.) The closed system for the first toss is generated by the event $\{HH, HT\}$ that the first toss is heads and the event $\{TH, TT\}$ that first toss is tails. Since the impossible and sure events belong to every closed system, there are four events in this closed system.

The smallest closed system is generated by the sure event Ω. It consists of the sure event and its complement, the impossible event. Since by definition one of the outcomes is sure to occur in any case, this closed system represents the state of no information about the result of the experiment. This may be a rather trivial example of a closed system, but it is an important special case: it represents the state of information before the experiment is conducted.

In the case when the closed system of events is infinite, it is customary to assume that unions and intersections of infinite sequences of events are also events. We shall always assume without further comment that this is the case. (In advanced probability this is the condition that the closed system of events be a "sigma-algebra" of events or a "sigma-field" of events.)

Example: Consider the target example, where the set of outcomes is a disk D. The largest closed system of events that is natural to consider consists of all subsets of D for which one can define a notion of area; this includes all rectangles and disks and many far more irregular subsets. However, one might only be interested in how far the outcome is from the center of the target. In that case the closed system might be smaller and contain only those events that can be defined in terms of the distance from the center. Typical events in this smaller closed system would be ring-shaped regions centered at the center point of the disk. Both of these examples are infinite, closed systems of events.

Probability measures The framework for probability is a set Ω of outcomes, a closed system $\mathcal{F}$ of events, and a *probability measure* P that assigns a *probability* $\mathbf{P}[A]$ satisfying $0 \leq \mathbf{P}[A] \leq 1$ to each event A in $\mathcal{F}$. This assignment is required to satisfy the mathematical properties that are specified below. The mathematical structure with all these properties is called a *probability space*.

The probability of the impossible event is $\mathbf{P}[\varnothing] = 0$. The probability of the sure event is $\mathbf{P}[\Omega] = 1$.

There are also rules for probabilities of combined events. The fundamental *additivity* rule for the union of exclusive events is

$$\mathbf{P}[A \textbf{ or } B] = \mathbf{P}[A] + \mathbf{P}[B].$$

In the case when the closed system of events is infinite, it is also standard to assume *countable additivity*, which means that the probability of the union of an infinite sequence of exclusive events is the infinite sum of the probabilities.

It follows from additivity that $\mathbf{P}[A] + \mathbf{P}[\textbf{not } A] = \mathbf{P}[A \textbf{ or not } A] = \mathbf{P}[\Omega] = 1$. Thus we have a rule for complementary events:

$$\mathbf{P}[\textbf{not } A] = 1 - \mathbf{P}[A].$$

It would be nice if there were a simple rule for the intersection of events. In general, there is no such rule. However, we define events A and B to be *independent* if

$$\mathbf{P}[A \textbf{ and } B] = \mathbf{P}[A] \cdot \mathbf{P}[B].$$

There is a product rule for independent events, but this is simply by the definition of independence.

The amount of dependence between events A and B is measured by how far the probability $\mathbf{P}[A \textbf{ and } B]$ differs from the product $\mathbf{P}[A] \cdot \mathbf{P}[B]$ associated with the individual events. Consider, for example, the situation when $\mathbf{P}[A] = 1/2$ and $\mathbf{P}[B] = 1/2$. If $\mathbf{P}[A \textbf{ and } B] = 1/4$, the events are independent. If $\mathbf{P}[A \textbf{ and } B] = 3/8$, the events are somewhat dependent. If $\mathbf{P}[A \textbf{ and } B] = 1/2$, the events are even more dependent; in fact, they are equivalent: the probability that one occurs and not the other is zero.

Sometimes one is considered to have partial information about the outcome of the experiment; for instance, the information about whether the outcome of the experiment is in B. The *conditional probability* is a revised prediction based on this partial knowledge. Assume that $\mathbf{P}[B] > 0$. The conditional probability corresponding to an outcome in B is defined to be $\mathbf{P}[A \mid B] = \mathbf{P}[A \textbf{ and } B]/\mathbf{P}[B]$. Consider again the case when $\mathbf{P}[A] = 1/2$ and $\mathbf{P}[B] = 1/2$. When A and B are dependent to the extent that $\mathbf{P}[A \textbf{ and } B] = 3/8$, the conditional probability of A given B is $\mathbf{P}[A \mid B] = 3/4$. This differs from the ordinary probability $\mathbf{P}[A] = 1/2$. The additional knowledge B has occurred makes A more likely to occur relative to this narrower context.

There is a special circumstance where the knowledge of the outcome B gives no useful information about whether A is more or less likely to occur. In this case, $\mathbf{P}[A \mid B] = \mathbf{P}[A]$. From the definition of conditional probability, this is the same as the condition that A and B be independent.

Probability theory is just a framework; it specifies properties of the probability measure, but it does not fix the measure for the particular experiment at hand. In order to define the probability measure, one must have some other theoretical information about the particular field of science, whether it is coin tossing or firing at a target.

It is possible to consider more than one probability measure defined on the same closed system of events. This just means that while the experiment outcomes are the same and the questions asked about the experiment are the same, there may be some different mechanisms for generating the results of the experiment. Thus one may bias the coin or change the marksman firing at the target. However, each such choice of mechanism defines a different probability space.

The following examples are both an illustration of these basic probability ideas and a preparation for some of the issues involving quantum mechanics and locality.

Example: Independent coin tossing A useful first example of a probability space is the space for an experiment consisting of two independent tosses of a fair coin. Think of the two tosses as taking place at very close to the same time but separated in space, say one on the left and one on the right. For instance, the left toss could take place in Seattle and the right toss in New York. The idea is that the timing and separation are to be arranged so that there is no possibility of direct communication between the two locations. It might then seem reasonable to assume that the two tosses are independent.

In the probability model the set $\Omega = \{HH, HT, TH, TT\}$ consists of the four possible outcomes involving heads or tails on the right or the left. The closed system $\mathcal{F}$ of events consists of all subsets of Ω. In this example the probability measure $\mathbf{P}$ is determined by its value on sets with one point:

$$\mathbf{P}[\{HH\}] = \mathbf{P}[\{TT\}] = \mathbf{P}[\{HT\}] = \mathbf{P}[\{TH\}] = \frac{1}{4}.$$

One can ask for the probability of the same result on the two tosses. By additivity, this is

$$\mathbf{P}[\{HH, TT\}] = \frac{1}{2}.$$

Similarly, the probability of a different result is

$$\mathbf{P}[\{HT, TH\}] = \frac{1}{2}.$$

These add to 1, as is required by additivity.

The events $\{HH, HT\}$ of having heads on the left toss and $\{HH, TH\}$ of having heads on the right toss are independent. The intersection of these events is the event $\{HH\}$, and

$$\mathbf{P}[\{HH\}] = \frac{1}{4} = \mathbf{P}[\{HH, HT\}] \cdot \mathbf{P}[\{HH, TH\}] = \frac{1}{2} \cdot \frac{1}{2}.$$

Certain events can be described in terms of what happens on the left without reference to what happens on the right. These form a closed system of events $\mathcal{F}_L$ generated by $\{HH, HT\}$ and $\{TH, TT\}$. Similarly, other events can be described in terms of what happens on the right without reference to what happens on the left. This closed system of events $\mathcal{F}_R$ is generated by $\{HH, TH\}$ and $\{HT, TT\}$. The independence of the tosses on the right and the left can be described by saying that every pair of events A and B such that A is in the closed system $\mathcal{F}_L$ and B is in the closed system $\mathcal{F}_R$ are independent.

Random variables A *random variable* is a rule that associates to each outcome a number. In other words, it is a number that depends on the outcome of the experiment.

In the coin-tossing example it is convenient to define a random variable σ^L that assumes the value 1 for the outcomes HH, HT and the value -1 for the outcomes TH, TT. This gives a score that only depends on the result of the left-hand coin toss. Similarly, there is a random variable σ^R that assumes the value 1 for the outcomes HH, TH and the value -1 for the outcomes HT, TT. This gives a score that only depends on the result of the right-hand coin toss.

It is customary in probability theory to use conditions on the values of random variables to define events. Thus, for example, the condition $\sigma^L = 1$ is a way of defining the event $\{HH, HT\}$, while the condition $\sigma^R = -1$ defines the event $\{HT, TT\}$.

A random variable is said to be *measurable* with respect to a closed system of events if the set of outcomes for which the random variable assumes a value (or range of values) is an event in the closed system. The random variable σ^L is measurable with respect to $\mathcal{F}_L$, while the random variable σ^R is measurable with respect to $\mathcal{F}_R$. This is because the value of σ^L depends only on what happens on the left, while the value of σ^R depends only on what happens on the right.

Example: Dependent coin tossing Here is another example of a probability space; it will occur later on in the context of quantum mechanics. Consider an experiment consisting of two coin tosses, again one on the left and one on the right. This time there is a conspiracy so that head or tail on one toss make it more likely that there will be the same head or tail on the other toss. Even though the probability of a head or tail on the left (or right) is $1/2$, the probability of the same outcome on both sides is $3/4$.

In this example the probability measure $\mathbf{P}$ is determined by its value on sets with one point:

$$\mathbf{P}[\{HH\}] = \mathbf{P}[\{TT\}] = \frac{3}{8}$$

and

$$\mathbf{P}[\{HT\}] = \mathbf{P}[\{TH\}] = \frac{1}{8}.$$

This is not the same as the assignment of probability for independent coin tossing.

Consider the random variables σ^L and σ^R that score 1 or -1 for heads or tails. One can ask for the probability of the event of the same result on the left and the right for the two tosses, that is, for the event $\sigma^L = \sigma^R$. By additivity, this is

$$\mathbf{P}[\sigma^L = \sigma^R] = \mathbf{P}[\{HH, TT\}] = \frac{3}{4}.$$

Similarly, the probability of a different result is

$$\mathbf{P}[\sigma^L = -\sigma^R] = \mathbf{P}[\{HT, TH\}] = \frac{1}{4}.$$

These add to 1, as is required by additivity.

The event $\sigma^L = 1$ of having heads on the toss on the left is $\{HH, HT\}$. The event $\sigma^R = 1$ of having heads on the toss on the right is $\{HH, TH\}$. These events are not independent. In fact, the probability of the event $\sigma^L = 1$ **and** $\sigma^R = 1$ is the probability of $\{HH\}$ which is

$$\mathbf{P}[\{HH\}] = \frac{3}{8} \neq \mathbf{P}[\{HH, HT\}] \cdot \mathbf{P}[\{HH, TH\}] = \frac{1}{2} \cdot \frac{1}{2}.$$

There are likely to be more HH outcomes in the long run than would be predicted by independence.

This can also be seen by calculating the conditional probability of a head on the left given a head on the right:

$$\begin{aligned} \mathbf{P}[\sigma^L &= 1 \mid \sigma^R = 1] \\ &= \mathbf{P}[\{HH, HT\} \mid \{HH, TH\}] \\ &= \mathbf{P}[\{HH\}]/\mathbf{P}[\{HH, TH\}] = \frac{\frac{3}{8}}{\frac{1}{2}} = \frac{3}{4}, \end{aligned}$$

which is not the same as the unconditional probability

$$\mathbf{P}[\sigma^L = 1] = \mathbf{P}[\{HH, HT\}] = \frac{1}{2}.$$

Imagine a person on the left and another person on the right, say one in Seattle and the other in New York. Each person will have only direct experience of events defined in terms of one of the coin tosses. Consider the two closed systems of events $\mathcal{F}_L$ and $\mathcal{F}_R$ such that events in $\mathcal{F}_L$ only depend on the outcome of the toss on the left and events in $\mathcal{F}_R$ only depend on the outcome of the toss on the right. The closed system $\mathcal{F}_L$ is generated by the events $\{HH, HT\}$ of heads on the left and $\{TH, TT\}$ of tails on the left. Similarly, $\mathcal{F}_R$ is generated by the events $\{HH, TH\}$ of heads on the right and $\{HT, TT\}$ of tails on the right. If we restrict the probability measure to $\mathcal{F}_L$, the only probability value for the events that generate $\mathcal{F}_L$ is $1/2$. Similarly, if one restricts the probability measure to $\mathcal{F}_R$, the only probability value for the events that generate $\mathcal{F}_R$ is $1/2$. Each of these restricted probability measures describes the toss of only one coin. Each of these defines a perfectly legitimate probability space, but the lack of independence of events on the left and on the right can be seen only by looking at events such as $\{HH\}$ that are neither in $\mathcal{F}_L$ or in $\mathcal{F}_R$. Such events are not in the direct experience of either person as long as they remain out of communication. However, if after each time the experiment is conducted the two people meet and compare their experiences, then they can answer questions about such events. If the experiment is repeated sufficiently many times so that the frequencies associated with the various events give a good estimate of the probabilities, then they can compare their results with the predictions of independence.

Enlarging a probability space Let Ω be a set of outcomes, $\mathcal{F}$ a closed system of events, and $\mathbf{P}$ a probability measure. Perhaps this turns out to be inadequate for the problem at hand. It may be possible to give a more complete description with another set Ω' of outcomes, another closed system $\mathcal{F}'$ of events, and another probability measure $\mathbf{P}'$.

Since the description is more detailed, there should be a rule that assigns to each outcome in Ω' a corresponding outcome in Ω. This is the rule that says to forget the extra detail.

This rule should also have the property that if A is an event in $\mathcal{F}$, then the set of outcomes in Ω' that the rule assigns to A is an event A' in $\mathcal{F}'$. This is to be thought of as the same event but described in the more detailed way. Of course, there may be other events in $\mathcal{F}'$ that use the extra detail as part of their description and do not arise from an event in $\mathcal{F}$.

Finally, if A is the event in the less complete description and A' is the corresponding event in the more complete description, then the probability $\mathbf{P}[A]$ should be the same as the probability $\mathbf{P}'[A']$.

An example of this is the set $\Omega = \{HH, HT, TH, TT\}$ of outcomes for the toss of two coins and the enlarged set for the situation where a third coin is tossed. We represent the result of the third coin toss by h for heads or t for tails. The enlarged set of outcomes is $\Omega' = \{HHh, HHt, HTh, HTt, THh, THt, TTh, TTt\}$. The rule that relates this larger set of outcomes to the original set of outcomes is to ignore the third toss. Thus the rule assigns to both HHh and HHt the outcome HH, to both HTh

and HTt the outcome HT, and so on. Consider an event in the smaller space, such as the event $\{HH, TT\}$ of having the same outcome for the two coins. The corresponding event in the larger space simply carries along the extra possibilities for third coin toss, so in the example it would be $\{HHh, HHt, TTh, TTt\}$. If the event in the first description has probability $3/4$, then the corresponding event in the more detailed description should also have probability $3/4$.

Example: Conditionally independent coin tossing One might want to go more deeply into the nature of the conspiracy. Here is a possible mechanism. There is a factory in Tucson that manufactures biased coins. The coins are cleverly engineered so that they have probability p for one side and probability q for the other side, where $p + q = 1$. The factory will take orders for whatever values of p and q are needed.

A preparation stage of the experiment takes place in Tucson. A fair coin is flipped, so that the probability of a head h is $1/2$ and the probability of a tail t is $1/2$. If the result is a head, coins are shipped to Seattle and New York that have probability p of heads and q of tails. If the result is a tail, coins are shipped to Seattle and New York that have probability q of heads and p of tails.

The final measurement stage of the experiment takes place in Seattle and New York simultaneously. The two coins that have been shipped there are flipped and the results are recorded. We assume that all three coin flips are independent. The probabilities for the total experiment are obtained by the multiplication rule and are

$$\mathbf{P}[\{HHh\}] = \mathbf{P}[\{TTt\}] = p^2\frac{1}{2}$$

and

$$\mathbf{P}[\{HHt\}] = \mathbf{P}[\{TTh\}] = q^2\frac{1}{2}$$

and

$$\mathbf{P}[\{HTh\}] = \mathbf{P}[\{THh\}] = \mathbf{P}[\{HTt\}] = \mathbf{P}[\{THt\}] = pq\frac{1}{2}.$$

From this we can calculate the probabilities for the final stage of the experiment alone.

$$\mathbf{P}[\{HHh, HHt\}] = \mathbf{P}[\{TTh, TTt\}] = \frac{p^2 + q^2}{2}$$

and

$$\mathbf{P}[\{HTh, HTt\}] = \mathbf{P}[\{THh, THt\}] = pq.$$

The dependent coin-tossing example can be recreated by having a standing order to the factory to produce p and q with $p + q = 1$ so that $pq = 1/8$. Then $p^2 = p - 1/8$ and $q^2 = q - 1/8$, so $(p^2 + q^2)/2 = 3/8$. The mathematicians employed at the factory are capable of working out the exact p- and q-values from these instructions: they are $(2 \pm \sqrt{2})/4$.

Each time the total experiment is conducted the entire process is repeated: a flip of a fair coin in Tucson that is used to select biased coins, a shipment of biased coins to the two coasts, and finally flips of biased coins in Seattle and New York. Certainly this elaborate process is sufficient to explain the dependence between the results in the two coastal cities.

Even though the results for the final stage are dependent, the dependence is due entirely to the work at the preparation stage. We can describe the result at this stage by a random variable τ that has the value 1 for heads at the preparation stage and -1 for tails at the preparation stage. The events at the preparation stage form a closed system $\tilde{\mathcal{F}}$ generated by the event $\tau = 1$ or $\{HHh, HTh, THh, TTh\}$ in which the coin at the first stage came up heads and the event $\tau = -1$ or $\{HHt, HTt, THt, TTt\}$ in which the coin at the first stage came up tails.

If we condition on the events in the closed system $\tilde{\mathcal{F}}$, we find that the events on the right and left at the second stage of the experiment are conditionally independent. For instance, we may compute the conditional probability of $\sigma^L = 1$ corresponding to heads on the left given $\tau = 1$, which is heads at the first stage. This is

$$\begin{aligned}\mathbf{P}[\sigma^L = 1 \mid \tau = 1]& \\ &= \mathbf{P}[\{HHh, HTh\}]/\mathbf{P}[\{HHh, HTh, THh, TTh\}] = p^2 + pq = p.\end{aligned}$$

The conditional probability of $\sigma^R = 1$ given $\tau = 1$ is also p. The conditional probability of $\sigma^L = 1$ **and** $\sigma^R = 1$ given $\tau = 1$ is

$$\mathbf{P}[\sigma^L = 1 \textbf{ and } \sigma^R = 1 \mid \tau = 1]$$

$$= \mathbf{P}[\{HHh\}]/\mathbf{P}[HHh, HTh, THh, TTh\}] = p^2,$$

which is the product of the two conditional probabilities.

One can formulate the conditional independence in a more general way in terms of the notion of conditional probability given the closed system $\tilde{\mathcal{F}}$. The closed system $\tilde{\mathcal{F}}$ is generated by the events $\tau = 1$ and $\tau = -1$. The conditional probability of an event A given the closed system $\tilde{\mathcal{F}}$ is defined to be the random variable $P[A \mid \tilde{\mathcal{F}}]$ that has value $P[A \mid \tau = 1]$ for all outcomes that belong to the event $\tau = 1$ and that has value $P[A \mid \tau = -1]$ for all outcomes that belong to the event $\tau = -1$. The idea is that the experiment is performed, but one has only partial information about the outcome, the information contained in the closed system $\tilde{\mathcal{F}}$. With this partial information about the result of the experiment, $P[A \mid \tilde{\mathcal{F}}]$ gives the best revised prediction of how likely A is to occur. Its value depends on the outcome, so it is a random variable.

The general statement of conditional independence given the preparation closed system is now strikingly simple: if A is an event in $\mathcal{F}_L$ and B is an event in $\mathcal{F}_R$, then $P[A \textbf{ and } B \mid \tilde{\mathcal{F}}] = P[A \mid \tilde{\mathcal{F}}]P[B \mid \tilde{\mathcal{F}}]$.

It might seem reasonable to formulate a general principle that dependence between simultaneous events at distant locations must be explained in this way in terms of prior preparation. This is the assumption of passive locality, and it will play a fundamental role in the discussion of alternatives to quantum mechanics.

Equivalent events In later discussions we will need to consider a property that guarantees that two events give the same probabilities in all contexts. We say that A and B are *equivalent* if

$$\mathbf{P}[A] = \mathbf{P}[A \textbf{ and } B] = \mathbf{P}[B].$$

It is easy to see that this is the same as the requirement that

$$\mathbf{P}[A \textbf{ and } \textbf{ not } B] = \mathbf{P}[B \textbf{ and } \textbf{ not } A)] = 0.$$

The significance of equivalence is that in all probability calculations B may be replaced by A (or vice versa) and the resulting probabilities will remain unchanged.

If an event is equivalent to the sure event, then it is said to occur *almost surely*. This means that even though it is not a logical necessity, it is a sure bet.

Example: Conditionally deterministic coin tossing We can consider a variant of the previous example in which the factory in Tucson manufactures two-headed and two-tailed coins. The experiment begins in Tucson with a first preparation stage at which a fair coin is flipped. If the coin comes up heads, then two-headed coins are shipped to both the Seattle and New York locations. If the coin comes up tails, then two-tailed coins are shipped to both places. When the two coins are simultaneously flipped, they must give the same result. This is a very strong dependence. This is in spite of the fact that the probability of heads or tails in either location is one-half.

Of course, the mechanism that allows this to happen is that whether heads or tails occurs at the second stage is entirely determined by the event at the preparation stage. Given the result of the first stage, the result of the second stage is purely deterministic.

The order to the factory for these coins is very simple: produce biased coins with probabilities p and q satisfying $p + q = 1$ and $pq = 0$. The only possibilities for p and q are 1 and 0, so this results in two-headed or two-tailed coins.

Since the preparation stage uses fair coins to decide which two-headed coins to use at the later experiment stage we have $\mathbf{P}[\{HHh\}] = 1/2$ and $\mathbf{P}[\{TTt\}] = 1/2$. All events that do not overlap with these have probability zero.

In this example we have the same closed systems as before. The closed system $\mathcal{F}_L$ on the left is generated by two events. One of these is the event $\sigma^L = 1$ in which heads appear on the left, which specifies the set $\{HHh, HTh, HHt, HTt\}$. The other is the event $\sigma^L = -1$ of tails on the left, which specifies $\{THh, TTh, THt, TTt\}$. Similarly, the closed system $\mathcal{F}_R$ on the right is generated by the heads event $\sigma^R = 1$ or $\{HHh, THh, HHt, THt\}$ and the tails event $\sigma^R = -1$ or $\{HTh, TTh, HTt, TTt\}$.

The probability assignments imply that the event $\sigma^L = 1$ of heads on the left and the event $\sigma^R = 1$ of heads on the right are equivalent to the event $\sigma^L = 1$ **and** $\sigma^R = 1$ or $\{HHh, HHt\}$ of heads for the two tosses. Also the event $\sigma^L = -1$ of tails on the left and the event $\sigma^R = -1$ of tails on the right are equivalent to the event $\sigma^L = -1$ **and** $\sigma^R = -1$ or $\{TTh, TTt\}$ of tails for the two tosses. We may summarize this by saying that $\sigma^L = \sigma^R$ almost surely. This is in spite of the fact that the events that generate $\mathcal{F}_L$ and $\mathcal{F}_R$ are quite random, with probability $1/2$ in each case.

The explanation for this coincidence is of course in the preparation. The preparation closed system is generated by the event $\tau = 1$ or $\{HHh, HTh, THh, TTh\}$ in which the preparation coin toss gives heads and the event $\tau = -1$ or $\{HHt, HTt, THt, TTt\}$ for which the preparation coin toss gives tails. The event $\sigma^L = 1$ of heads on the left and the event $\sigma^R = 1$ of heads on the right and the event $\tau = 1$ of the preparation toss giving heads are all equivalent to the event $\{HHh\}$ of heads for all three tosses. Also the event, $\sigma^L = -1$ of tails on the left and the event $\sigma^R = -1$ of tails on the right and the event $\tau = -1$ of the preparation toss giving tails are all equivalent to the event $\{TTt\}$ of tails for all three tosses. The fact that $\sigma^L = \sigma^R$ almost surely is explained by the fact that $\sigma^L = \tau = \sigma^R$ almost surely.

Again there is a general principle: such a perfect coincidence or equivalence between simultaneous distant random events must arise because they are both determined by some random event in the past. This is a deterministic version of passive locality.

Example: A target Here is another example of a probability space, this time with infinitely many outcomes. This one will be useful later on in constructing a hidden-variable model of spin. Consider a circular target, represented by a disk of radius 1. This disk D is the set of outcomes. The closed system $\mathcal{F}$ of events includes all subsets of the disk that have a well-defined area. We take the probability measure to be the *uniform probability* $\bar{\mathbf{P}}$. The uniform probability of a subset A of the disk is defined to be proportional to its area:

$$\bar{\mathbf{P}}[A] = \frac{1}{\pi}\text{Area}[A].$$

Dividing by π makes the definition satisfy the property that the probability of the entire disk D is equal to 1. The uniform probability is appropriate to describe the results of a very inaccurate marksman firing at the target D.

In this type of probability space each set that has only one point has probability zero. (The area of a single point is zero.) This is in spite of the fact that one is sure to hit some point in the target!

Also, each set that is a line segment or a circle has probability zero. (The area of a set with nonzero length but zero width is zero.) This may seem upsetting at first, but it allows flexibility in specifying events. Say that one is interested in hitting near the center of the target, say within a distance a. The distance of the hit from the center is a random variable r. One could consider the event of a successful hit to be the set A of points with radius $r < a$. Then the probability $\bar{\mathbf{P}}[A] = a^2$. Instead one could consider the event to be the set A' of points with radius $r \leq a$. The probability $\bar{\mathbf{P}}[A'] = a^2$ is the same. Indeed, A and A' are equivalent events. The reason is that the point where the target is hit will almost surely not be on the circle of radius a.

PASSIVE LOCALITY

The evaluation of alternatives to quantum mechanics involves locality arguments. Locality is a constraint on what can happen simultaneously at distant locations. Active locality rules out instantaneous direct intervention. Passive locality rules out unexplained dependence of random events at the distant locations. It is possible to have active locality without passive locality, since unexplained dependence cannot necessarily be used to intervene or signal. One might not even notice the dependence until much later!

In this section we concentrate on the notion of passive locality. There are a large number of alternative formulations and variant terminologies in discussions of locality. (For instance, active locality and passive locality are sometimes called "parameter independence" and "outcome independence.") In this treatment we use the terminology of Nelson (see [1] with corrections in [2]). In his terminology, *passive locality* is the condition that for simultaneous but widely separated random events there must be prior preparation events such that the separated events are conditionally independent given the preparation. (This condition is known in the philosophy literature [3] as the principle of *common cause*.)

Consider random events widely separated in space but taking place almost simultaneously in such a way that there is no possibility of communication in either direction that could affect the outcome. In a probability description these events could be independent or they could be dependent. The assumption of passive locality is that if these events are dependent, then they must be so only because there was a component of randomness introduced by some prior preparation. Given a complete description of the results of the preparation, any residual randomness is independent. The example to keep in mind is dependent coin tossing that becomes conditionally independent given the results of the preparation stage.

Passive locality is a condition involving not only probability but also space and time. To use it one needs to relate it to some other condition that refers to these same concepts. In the proof of Bell's second theorem below, the two related conditions are active and passive locality.

The version of the proof given below uses only an assumption that is a consequence of passive locality. We shall refer to this as *deterministic passive locality*. The observation that deterministic passive locality is a consequence of passive locality is attributed by Redhead [4] to Suppes and Zanotti. The technical note ahead provides a proof.

Consider random events that take place widely separated in space but almost simultaneously. Deterministic passive locality requires that if these events are equivalent, in the sense that with probability 1 they always happen or do not happen together, then this must be because they are equivalent to some random events at a prior preparation stage. Given a complete description of the results of the preparation, the subsequent events are determined. All the randomness must be in the preparation. Here the example is the two coins in widely separated locations that always give the same result, since they were chosen by chance at the preparation stage to be either both two-headed coins or both two-tailed coins.

Deterministic passive locality has a mathematical formulation. The situation of interest is a probability experiment that takes place almost simultaneously at two widely separated locations, say on the left and on the right. There are a set of outcomes, a closed system of events, and a probability measure. Some of the events take place at the left; others at the right. We shall refer to these as the closed system of *left events* and the closed system of *right events*.

The assumption is that this probability space may be enlarged to a probability space involving a prior preparation stage such that certain properties are satisfied. Thus there are a set Ω of outcomes for the enlarged experiment, a closed system $\mathcal{F}$ of events for the enlarged experiment, and a probability measure P for the enlarged experiment defined on the events belonging to $\mathcal{F}$. The enlarged experiment has corresponding closed systems $\mathcal{F}_L$ and $\mathcal{F}_R$ of left and right events that are contained in the closed system $\mathcal{F}$.

The experiment takes place in at least two stages. The first stage is the preparation stage. Events at this stage are events that take place at the time of the preparation. We shall refer to this closed system of *preparation events* as $\tilde{\mathcal{F}}$. The closed system $\tilde{\mathcal{F}}$ is also contained in the closed system $\mathcal{F}$ for the entire experiment.

The second stage of the experiment is the measurement stage. The left closed system $\mathcal{F}_L$ and the right closed system $\mathcal{F}_R$ are associated with this stage. The events in these closed systems also belong to the closed system $\mathcal{F}$ for the entire experiment. There may also be events that belong neither to $\mathcal{F}_L$ or to $\mathcal{F}_R$.

Here is the assumption of deterministic passive locality. Let A be an event in $\mathcal{F}_L$ and A' be an event in $\mathcal{F}_R$. Assume that A and A' are equivalent with respect to $\mathbf{P}$. Then there is an event $\tilde{A}$ in $\tilde{\mathcal{F}}$ such that A and A' are each equivalent to $\tilde{A}$ with respect to $\mathbf{P}$.

The meaning of the property that A and A' are equivalent is that the event that A happens is determined by A' happening, and vice versa. Since the events A and A' are determined simultaneously in very distant regions, then, in the absence of nonlocal instantaneous influence, the events must coincide because they were set up to do so in the past. The way they were set up is given by the event $\tilde{A}$ defined at the preparation stage.

Note that this past event $\tilde{A}$ may be random; the deterministic passive locality assumption is only that no new randomness can be introduced at the measurement stage without spoiling the perfect coincidence of the events A and A'. The reason is that the experimental apparatus consists of two remote parts, and randomness introduced in one part should not be able to perfectly compensate for randomness introduced in the other part, at least not without some sort of conspiracy between the two parts.

What is the status of the assumption of passive locality or of deterministic passive locality? Are they satisfied by physical theories? Are they required by physical theories? These questions do not have obvious answers. It seems difficult to imagine a physical theory involving probability that does not satisfy passive locality, but this does not mean that it could not exist. In one view quantum mechanics is such a theory. Still, many physicists would regard it as a plausible constraint. The reader can form his or her own judgment.

Technical note:

The condition of passive locality is that events in $\mathcal{F}_L$ and $\mathcal{F}_R$ are conditionally independent given $\tilde{\mathcal{F}}$. We show that passive locality implies deterministic passive locality. This note uses concepts from advanced probability and is intended for the expert.

Passive locality says any randomness introduced at the measurement stage on the left must be independent of any randomness introduced at the measurement stage on the right. Deterministic passive locality says that if the distant events are equivalent, then there cannot be any randomness introduced at the measurement stage. Independent random variation would destroy the perfect equivalence of two distant events.

The conditional probability of an event E given $\tilde{\mathcal{F}}$ is a random variable $\mathbf{P}[E \mid \tilde{\mathcal{F}}]$ with values between zero and one. (Here the closed system of events $\tilde{\mathcal{F}}$ is assumed to be what is technically known as a sigma-algebra or a sigma-field.) It is characterized by the following three properties:

1. The random variable $\mathbf{P}[E \mid \tilde{\mathcal{F}}]$ is measurable with respect to $\tilde{\mathcal{F}}$.
2. If G is in $\tilde{\mathcal{F}}$, then $\mathbf{P}[E \textbf{ and } G \mid \tilde{\mathcal{F}}] = \mathbf{P}[E \mid \tilde{\mathcal{F}}]$ for outcomes in G, and $\mathbf{P}[E \textbf{ and } G \mid \tilde{\mathcal{F}}] = 0$ for outcomes not in G.
3. The probability $\mathbf{P}[E]$ may be computed by taking the expectation of the conditional probability:

$$\mathbf{P}[E] = \mathbf{E}[\mathbf{P}[E \mid \tilde{\mathcal{F}}]].$$

The intuitive interpretation of the conditional probability corresponding to a given outcome is as the revised probability that E will occur given the extra information about which events in $\tilde{\mathcal{F}}$ occur for this outcome.

Now assume that A in $\mathcal{F}_L$ and A' in $\mathcal{F}_R$ are equivalent. This means that $\mathbf{P}[A] = \mathbf{P}[A \textbf{ and } A'] = \mathbf{P}[A']$.

Passive locality says that the conditional probabilities satisfy the independence property

$$\mathbf{P}[A \textbf{ and } A' \mid \tilde{\mathcal{F}}] = \mathbf{P}[A \mid \tilde{\mathcal{F}}] \cdot \mathbf{P}[A' \mid \tilde{\mathcal{F}}]$$

almost surely.

Next observe that

$$\mathbf{P}[A \textbf{ and } A'] = \mathbf{E}[\mathbf{P}[A \textbf{ and } A' \mid \tilde{\mathcal{F}}]]$$

and

$$\mathbf{P}[A] = \mathbf{E}[\mathbf{P}[A \mid \tilde{\mathcal{F}}]].$$

Now the two random variables satisfy $\mathbf{P}[A \textbf{ and } A' \mid \tilde{\mathcal{F}}] \leq \mathbf{P}[A \mid \tilde{\mathcal{F}}]$ almost surely. Since by the observation and the equivalence of A and A' their expectations are equal, we may conclude that $\mathbf{P}[A \textbf{ and } A' \mid \tilde{\mathcal{F}}] = \mathbf{P}[A \mid \tilde{\mathcal{F}}]$ almost surely. Similarly, $\mathbf{P}[A \textbf{ and } A' \mid \tilde{\mathcal{F}}] = \mathbf{P}[A' \mid \tilde{\mathcal{F}}]$ almost surely.

It follows that the independence property may be written to say that

$$\mathbf{P}[A \mid \tilde{\mathcal{F}}] = \mathbf{P}[A \mid \tilde{\mathcal{F}}]^2$$

almost surely. This last equation implies that the conditional probability $\mathbf{P}[A \mid \tilde{\mathcal{F}}]$ can only have values zero and one, almost surely. We denote the event where this random variable has the value one by $\tilde{A}$. Since the conditional probability is measurable with respect to $\tilde{\mathcal{F}}$, this event must be in $\tilde{\mathcal{F}}$.

The computation of the probability of A in terms of the expectation of the conditional probability gives

$$\mathbf{P}[A] = \mathbf{P}[\tilde{A}].$$

Since $\tilde{A}$ is in $\tilde{\mathcal{F}}$, it legitimate to compute the probability of A **and** $\tilde{A}$ in the same way to obtain

$$\mathbf{P}[A \textbf{ and } \tilde{A}] = \mathbf{P}[\tilde{A}].$$

These last two results establish the conclusion that A is equivalent to $\tilde{A}$ in $\tilde{\mathcal{F}}$. Similarly, A' is equivalent to the same $\tilde{A}$. This shows that passive locality implies deterministic passive locality.

QUANTUM MECHANICS

Quantum mechanics is an extraordinarily successful theory on many levels; it is perhaps the deepest insight that we have into Nature. It works on many levels: elementary particles, nuclei, atoms, molecules, and bulk matter. Quantum mechanics together with known forces comes close to explaining the entire physical world. The theory of elementary particles is not yet complete, and gravitation stands somewhat apart from the rest of physical theory, but on the whole we have a reasonably good picture of ordinary matter.

In particular, quantum mechanics plus electrostatics explains most properties of everyday objects: density, compressibility, conductivity, color, chemical reactivity, and so on. Why is a table solid? Because the random jiggling of electrons predicted by quantum mechanics balances the electrostatic attraction between electrons and nucleus. Why is grass green? Because the quantum oscillations of the molecules resonate with green light.

Quantum mechanics is formulated in highly abstract terms. The quantum mechanical "state" expresses in an indirect way various probabilities involving particle observables. The mathematical apparatus of quantum mechanics does not correspond to any simple picture of Nature; its justification is its success in making predictions.

Mathematics of quantum mechanics The following is not intended to be a complete description of quantum mechanics. The intention is to say the minimum that is required and to say nothing false.

Probability theory is formulated in terms of a set of *outcomes*, a closed system of *events*, and a *probability measure*. Experimental quantities are represented by *random variables*, which are functions on the set of outcomes.

There is no single set of outcomes in quantum mechanics. There are, however, *quantum events* corresponding to various possible measurements. The role of the probability measure is played by the quantum *state*. Experimental quantities corresponding to various possible measurements are represented by *quantum random variables*, more commonly called *observables*. (In many expositions of quantum mechanics, a state is represented by a "wave function," and an observable corresponds mathematically to an "operator.")

The difference between probability theory and quantum mechanics is that in quantum mechanics there is no single notion of outcome of the experiment. A measurement of one observable ordinarily precludes a simultaneous measurement of another observable. There is no objective outcome for all observables.

Often the conceptual difficulties of quantum mechanics are illustrated by the case of spin measurements. When the spin of a particle of spin $1/2$ is measured there are only two possible values; it is like a coin toss. For two such particles there are only four possible values that result from a measurement; this is like an experiment with two coin tosses. In the following we formulate quantum mechanics only for the special case of systems for which measurements can have only a finite number N of possible outcomes. This eliminates certain mathematical subtleties and is sufficient for these cases of spin measurements, where $N = 2$ or $N = 4$.

The mathematical framework for quantum mechanics is formulated in terms of a set (the set of all states) and a function that associates to each pair u and v of states a number $p(u, v)$. The numbers $p(u, v)$ are called *transition probabilities*.

The set of states and the transition probability function are mathematical objects that do not correspond to any physical picture. One can sometimes attempt to picture them as geometrical objects. Thus, for instance, for a spin $1/2$ particle, the set of states is an ordinary two-dimensional sphere and the function $p(u, v) = \cos^2(\theta/2)$, where θ is the angle between the points u and v on the sphere. For a pair of spin $1/2$ particles, the set of states is a much more complicated six-dimensional object, so it is difficult to picture except by analogy. However, in every case the ultimate role of these mathematical objects is simply to predict probabilities for certain experiments.

In quantum mechanics the transition probability function always satisfies the following properties:

Property 1. *The transition probability is symmetric: for each pair of states u and v, we have $p(u, v) = p(v, u)$.*

Property 2. *For each pair of states u and v the transition probability $p(u, v)$ satisfies $0 \leq p(u, v) \leq 1$. States have transition probability 1 if and only if they are the same state: $p(u, v) = 1$ is equivalent to $u = v$.*

We say that two states u and v are *opposite* if $p(u, v) = 0$. A set of states is said to be *opposite* if every pair of distinct states is opposite. A opposite set is said to be *maximal opposite* if it is not contained in any larger opposite set. (This use of the term "opposite" to describe a relation between states is not standard, but it is appropriate to a geometric description of the set of all states as a "projective space.")

Property 3. *Every opposite set is contained in a maximal opposite set. Each such maximal opposite set contains the same number N of states.*

Property 4. *Let $e_1, e_2, \ldots, e_N$ be a maximal opposite set. Then for every state u,*

$$p(u, e_1) + p(u, e_2) + \cdots + p(u, e_N) = 1.$$

The analog in quantum mechanics of the specification of a probability measure is the specification of a *state u*. In the mathematical formulation of quantum mechanics, a *possible measurement* is a maximal opposite set $\{e_1, e_2, \ldots, e_N\}$. For each state and each possible measurement, there is a corresponding probability space. However, different possible measurements give rise to different probability spaces.

This concept of possible measurement does not give much of an idea as to how to identify what is being measured or how to carry out the measurement. For the moment it is just part of the mathematical apparatus. Clearly, additional information is needed to use this for a particular situation.

For each possible measurement we have a set of outcomes $\Omega_e = \{e_1, e_2, \ldots, e_N\}$. We also have a closed system of events that consists of all subsets of Ω_e. The state u determines a probability measure $\mathbf{P}_u$ defined on subsets of Ω_e such that

$$\mathbf{P}_u[\{e_i\}] = p(u, e_i).$$

The role of the state u is to determine for each possible measurement a corresponding probability measure. We refer to the specification of a possible measurement and an event associated to this possible measurement as a *quantum event*.

Consider a state u and a possible measurement represented by a maximal opposite set $\{e_1, e_2, \ldots, e_N\}$. A random variable X is a function that assigns to the outcomes $e_1, e_2, \ldots, e_N$ in Ω_e corresponding values $x_1, x_2, \ldots, x_N$. The probability $\mathbf{P}_u[X = x]$ that the random variable X has value x is the sum of $p(u, e_i)$ over all i with $x_i = x$. We refer to the specification of a possible measurement and a random variable associated to this possible measurement as a *quantum random variable*. In standard quantum mechanics terminology, a quantum random variable is called an *observable*.

Quantum mechanics and probability How is quantum mechanics related to probability? The specification of a state u is analogous to the specification of a probability measure. However, corresponding to the various possible measurements, there are various probability spaces, each with its own probability measure. A subset of the set of outcomes of one of these possible measurements is a quantum event. The complement of a quantum event is another quantum event corresponding to the same possible measurement. However, in general, the union and intersection of two quantum events is not defined if the quantum events are associated with different possible measurements.

A random variable defined on the outcomes of one of these possible measurements is a quantum random variable, or in the language of quantum mechanics, an observable. The peculiarity of quantum mechanics is that different observables can be random variables on different probability spaces.

The usual interpretation of this situation is that the act of measuring one observable precludes in principle the act of observing an observable associated with another possible measurement. The idea of an overall outcome of an experiment that determines the values of all observables is meaningless. There is no reason that one should be able to consider results involving two observables in connection with a single trial of an experiment. (Of course, in multiple trials there is ample opportunity for measuring probabilities associated with various observables.)

Consider a situation when the state u is fixed. For each possible experiment one can compute probabilities associated with the maximal opposite set. Thus one can compute probabilities associated with observables corresponding to each possible experiment. However, quantum mechanics gives no definition of joint probabilities (except in certain special situations). Consider the typical situation in which X is an observable corresponding to one possible experiment with outcomes in $\Omega_e = \{e_1, e_2, \ldots, e_N\}$ and Y is an observable corresponding to another possible experiment with outcomes in $\Omega_f = \{f_1, \ldots, f_N\}$. The probability $\mathbf{P}_u[X = x]$ that the random variable X has value x is the sum of $p(u, e_i)$ over all i with $x_i = x$. Similarly, the probability $\mathbf{P}_u[Y = y]$ is the sum of $p(u, f_j)$ over all j such that $y_j = y$. There are two different probability spaces and two different probability measures. There is no meaning given to events such as $X = x$ **and** $Y = y$, and so there are also no corresponding probabilities.

What happens after the measurement has taken place? The name "transition probability" for $p(u, e_i)$ suggests that when a measurement is made, the state makes a transition from the original state u to the state corresponding to the actual outcome e_i. However, it may be that a more detailed theory of the quantum measurement process is needed to describe state change during measurement.

The formulation of quantum mechanics does not attempt to give a picture of a real world apart from predictions of measurement. A sufficiently well-isolated system has a state that would predict the probabilities of the results of various measurements; until these measurements are made, the usual observables have no definite values. This is not because the values are unknown—the analysis purports to show that there can be no meaning to these quantities independent of measurement. So there is a sense in which quantum mechanics does not give a realistic view of the world.

Technical note

This note is intended for people familiar with other treatments of quantum mechanics. The usual mathematical formulation of quantum mechanics is in terms of vectors that can be added and multiplied by complex numbers. These vectors form what is called an N- dimensional *complex vector space*. The vectors are equipped with a suitable notion of length and orthogonality; in the usual terminology it is said that they form a *Hilbert space*. Quantum mechanical states are represented by vectors of length 1, except that two such vectors that are complex multiples of each other represent the same vector. So the space of states is actually an $(N-1)$-dimensional *complex projective space*. (Since each complex number corresponds to a pair of real numbers, the dimension as measured by the number of real parameters is $2(N-1) = 2N-2$.)

In this more specific mathematical framework, the condition that two states be opposite is that they are represented by two vectors that are orthogonal. (Thus "opposite" is the technical term appropriate to the view that states belong to projective space, while "orthogonal" is natural in the vector space framework.) Let χ and χ' be two orthogonal vectors of length 1. They define opposite states e and e'. A combination

$$\psi = \cos(\theta/2)\chi + \sin(\theta/2)e^{i\phi}\chi'$$

defines a state u called a *superposition* of e and e'. The set of superpositions of e and e' forms a sphere, where the angle from the north pole is measured by θ, and the longitude angle measured from some arbitrary meridian is ϕ. So complex projective space is constructed from spheres.

If u and e are states represented by vectors ψ and χ, then their transition probability $p(u, e)$ is given in terms of the complex inner product $\langle\psi, \chi\rangle$ of ψ and χ by

$$p(u, e) = |\langle\psi, \chi\rangle|^2 = \cos^2(\theta/2).$$

Quantum mechanics is the geometry of spheres.

A SPIN 1/2 QUANTUM SYSTEM

The simplest quantum system is the one describing the spin states of a spin 1/2 particle, such as an electron or a proton. The idea is that for each direction in space, there is a variable called the spin. The remarkable fact is that each of these spin variables can have only two possible values.

The spin of a particle is a property of the particle that is usually measured in units of the standard quantum mechanical measure of angular momentum (the rationalized Planck's constant). There can be particles of spin 0, spin 1/2, spin 1, and so on. Electrons, protons, and neutrons all have spin 1/2. Since we shall only be dealing with spin 1/2 particles, we shall measure spins in units of half the standard unit, so that the possible values will be multiples of 1. This gets rid of some irritating fractions.

How does spin correspond to an actual laboratory experiment? The idea is that the spinning particle is in motion and passes through an experimental apparatus. This apparatus involves a magnetic field whose gradient is in the direction e. It is an experimental fact that the particle is deflected either in the direction e or in the opposite direction e'. The final measurement is actually a measurement of the position of the deflected particle when it arrives at a detection apparatus; however, it is interpreted as the result of an interaction between the magnetic field gradient and the spin of the particle. When the experiment is repeated many times, it is seen that the proportion of time that it is deflected in the direction e is given by a certain probability. We think of this as being a measurement of a quantum random variable σ_e and the probability as being the probability of the quantum event $\sigma_e = 1$.

In each run of the experiment, the decision of what to measure is determined by the direction of the magnetic field gradient. There are, in principle, infinitely many possible directions. Making the measurement corresponding to one direction precludes a simultaneous measurement corresponding to another direction. It is possible to make a number of runs of the experiment in each of several directions of the magnetic field gradient. In that way one can get the individual probabilities corresponding to the various directions.

One can think of this experiment as a huge family of coin-toss experiments, one corresponding to each direction of space. However, in conducting the experiment, one chooses just one of these directions and makes the coin toss for this direction alone.

The experiment is described by quantum mechanics in the following way. The states of such a system correspond to the points on a sphere. Each point corresponds to a spin direction. The angle between two points on a sphere is measured along the shortest arc of a great circle joining the two points. This angle is always between 0 degrees and 180 degrees. The transition probability between two states u and v is

$$p(u, v) = \cos^2(\theta/2),$$

where θ is the angle between u and v. This expression results from a deep analysis of the symmetries of spinning particles; we shall just take it as a given.

A possible measurement is determined by a pair of opposite points e and e'. The values of the spin observable σ_e corresponding to these points are 1 and -1. Note that if θ is the angle between u and e, then

$$p(u, e) = \cos^2(\theta/2),$$

while

$$p(u, e') = \sin^2(\theta/2).$$

These are the probabilities associated with values of the spin observables. Consider a state corresponding to the u direction on the sphere. For the observable σ_e associated with the e direction, the probability space of outcomes is $\Omega_e = \{e, e'\}$. When a measurement is made of the component of spin in the e direction, the probability that the component is 1 is

$$\mathbf{P}_u[\sigma_e = 1] = \mathbf{P}_u[\{e\}] = \cos^2(\theta/2),$$

where θ is the angle between e and u. Similarly, the probability that the component is -1 is

$$\mathbf{P}_u[\sigma_e = -1] = \mathbf{P}_u[\{e'\}] = \sin^2(\theta/2).$$

To get the special flavor of quantum mechanics, we need to compare different spin observables. There is one situation that needs special consideration. Let e' be the point on the sphere opposite to e. Then when the spin component σ_e is ± 1, the spin component $\sigma_{e'} = \mp 1$. Thus the values of σ_e and $\sigma_{e'}$ are negatives of each other.

The general situation is quite different, and it illustrates the distinctive feature of quantum mechanics. Two points e and f that are not equal or opposite do not belong in a single possible measurement. Thus if e and f are points that are not equal or opposite, then the probability spaces $\Omega_e = \{e, e'\}$ and $\Omega_f = \{f, f'\}$ are distinct. The observables σ_e and σ_f are not jointly observable; they are random variables defined on two different probability spaces. There is no quantum event $\sigma_e = 1$ **and** $\sigma_f = 1$. The probabilities are associated with each observable separately and not with the whole collection.

Note on geometry and trigonometry At least some mathematicians use a consistent terminology in geometry, in which a *circle* is a curve that bounds a planar region called a *disk*. Similarly, a *sphere* is a surface that bounds a solid region called a *ball*.

Consider a circle of radius 1. The straight line distance between two points on the circle at an angle θ is $d_\theta = 2\sin(\theta/2)$. The corresponding transition probability is related to this distance by $p_\theta = 1 - (d_\theta/2)^2 = \cos^2(\theta/2)$, so it has at least some relation to geometry.

Calculations of such transition probabilities involve values of the cosine function for angles between 0 degrees and 180 degrees. The cosines of angles of 0, 30, 45, 60, and 90 degrees are 1, $\sqrt{3}/2$, $\sqrt{2}/2$, $1/2$, and 0. The cosines of other angles may be computed from the identity $\cos(\theta) = -\cos(180° - \theta)$. The most useful identities are the ones that give alternate forms for the probabilities of spin in two opposite directions:

$$\cos^2(\theta/2) = \frac{1}{2}(1 + \cos(\theta))$$

and

$$\sin^2(\theta/2) = \frac{1}{2}(1 - \cos(\theta)).$$

When we add these, we get

$$\cos^2(\theta/2) + \sin^2(\theta/2) = 1,$$

which is the equation that says that the total probability is 1. When we subtract them, we obtain

$$\cos^2(\theta/2) - \sin^2(\theta/2) = \cos(\theta),$$

which is the mathematical expectation obtained by weighting the possible values ± 1 by the corresponding probabilities.

It is helpful to have a table of the transition probabilities $\cos^2(\theta/2)$. For angles of 0, 60, 90, 120, 180 degrees, the probabilities are 1, 3/4, 1/2, 1/4, and 0, respectively.

PROBABILITY MODEL OF SPIN 1/2

The spin 1/2 particle is a rather exceptional quantum system. It is possible in this one case to make a natural model in which there is a well-defined outcome of a probability experiment that determines the values of all the observables. The picture is simple. The outcome is a point on the sphere, and the random variable σ_e corresponding to direction e has the value 1 if the outcome is in the hemisphere with e as the north pole and has the value -1 if the outcome is in the opposite hemisphere.

This can still be thought of as something like a family of coin-toss experiments, one for each direction of space. However, once the experiment results in a particular point on the sphere, the results of all the coin tosses are simultaneously determined by this one result: they give heads on the hemisphere corresponding to the point and tails on the opposite hemisphere.

This model was presented in the paper by Kochen and Specker [5]. The quantum mechanical state is a point u on the sphere of radius 1. However, now there is a probability space corresponding to this state. The set Ω of all possible outcomes is the sphere. The closed system $\mathcal{F}$ of events consists of all subsets of Ω for which a reasonable notion of surface area may be defined; thus it includes practically every imaginable subset. The probability measure P_u assigns a probability to each event. This measure has the property that the probability of the event that the outcome is in the hemisphere determined by u is 1.

Here is the precise definition of the probability measure corresponding to the state u. Recall that u is a point on the sphere of radius 1. Consider the flat disk D_u of radius 1 passing through the center of the sphere and perpendicular to the line from the center of the sphere to u. For each point in D_u there is a projected point in the sphere Ω that lies in the hemisphere of the sphere containing u. This is obtained by projecting in a straight line perpendicular to the disk. This projection operation will be used to define the probability measure corresponding to u.

We already have the uniform probability measure $\bar{\mathbf{P}}$ defined on subsets of the disk D_u, for which the probability is proportional to the area. This defines the desired probability measure $\mathbf{P}_u$ as follows. The probability of the event A (a subset of the sphere Ω) is

$$\mathbf{P}_u[A] = \bar{\mathbf{P}}[A_u],$$

where $\bar{\mathbf{P}}$ is the uniform probability on D_u, and A_u is the set of points on the disk D_u that project onto A.

One can think of the experiment in a very concrete way as follows. A marksman is located very far from the sphere in the direction indicated by u. He fires a shot that penetrates the sphere Ω and then hits the disk D_u at a random point. The point where it penetrates the sphere is the outcome. The probability $\mathbf{P}[A]$ is the theoretical prediction of the chance that such an outcome is in the subset A of the sphere. Since the shot arrives at the sphere from the side indicated by u, the probability that the outcome is in the hemisphere determined by u is 1.

The spin observables σ_e of quantum mechanics corresponding to various directions e have 1 and -1 as their only possible values. In order to fit them into the probability model, they must take the form of random variables defined on the set Ω of all outcomes. The definition is that $\sigma_e = 1$ when ω is in the hemisphere determined by e, and $\sigma_e = -1$ when ω is not in this hemisphere.

It is not too difficult to compute the probabilities and see that they give the same result as quantum mechanics. The calculation is sketched in the note below. The conclusion is that

$$\mathbf{P}_u[\sigma_e = 1] = \cos^2(\theta/2),$$

where θ is the angle between e and u. This agrees with the answer computed by the quantum mechanics of spin $1/2$.

However, the probability model gives us much more. For example, we may compute joint probabilities involving σ_e and σ_f. The probability that these are both 1 is $\mathbf{P}_u[\sigma_e = 1 \textbf{ and } \sigma_f = 1]$, which is just the probability associated with the region A common to the two hemispheres associated with e and with f.

All this is a reflection of the fact that we have a very concrete model for the result of an experiment involving the spin of a single spin $1/2$ particle in a state corresponding to direction u. When the experiment is conducted, the outcome is some random direction in the unit sphere in the same hemisphere as u. This outcome is an intrinsic property of the particular spinning system; it has no counterpart in quantum mechanics.

The spin random variable σ_e corresponding to the direction e has the value 1 or -1 depending on whether the outcome is also in the same hemisphere as e. Thus it only depends on whether the outcome is "seen" from the same side as e. So the values of all such random variables depend on the outcome. Of course, on different runs of the experiment one gets different outcomes; the long-run behavior is predicted by the probability measure P_u.

Since the random outcome is always in the hemisphere seen from the direction u, the spin random variable σ_u in the direction u given by the state always has the value 1. So in this state the spin in this particular direction has a definite nonrandom value. Similarly, the spin in the opposite direction u' is sure to have the value -1. The spin in any other direction e will sometimes have the value 1 and sometimes the value -1, depending on whether the outcome happens to be visible from this direction or not. The hitherto mysterious quantum mechanical probabilities $\cos^2(\theta/2)$ and $\sin^2(\theta/2)$ simply give the chances for these two events.

Kochen and Specker do not attempt to give a more detailed account of the mechanism by which the discrete values of σ_e arise in an actual experiment; their point is that nothing in probability theory rules out the existence of such a mechanism.

Technical note

Here is a calculation of the probability when the state is determined by u of the event that $\sigma_e = 1$. The result to be proved is that this probability is $(1/2) + (1/2)\cos(\theta) = \cos^2(\theta/2)$, where θ is the angle between e and u. Recall that u and e are each points on the sphere. The probability measure on the sphere determined by u is obtained from the uniform probability measure on the disk D_u passing through the center of the sphere perpendicular to the line from the center to u. It is the projection of this measure on the sphere in the direction of u. The event that $\sigma_e = 1$ corresponds to the hemisphere determined by e. Thus the probability that $\sigma_e = 1$ is the uniform probability of the part of D_u that projects onto the intersection of the hemispheres determined by u and by e.

It is enough to do the calculation in the case when u and e lie in the same hemisphere at an angle θ on the sphere. The other case, when u and e lie in opposite hemispheres, may be reduced to the first case by noting that the event $\sigma_e = 1$ is the complement of the event $\sigma_{e'} = 1$, where e' is opposite to e.

Consider the disk D_e passing through the center of the sphere perpendicular to the line from the center to e. The intersection of D_e with D_u is a line segment that divides the disk D_u into two halves. All of one half projects onto the intersection of the hemispheres, but only a portion of the other half. The probability associated with the first half is $1/2$. The portion of the other half may also be described as the projection at angle θ of half of the disk D_e onto part of half of the disk D_u. Since projection at an angle θ multiplies corresponding areas by a factor $\cos(\theta)$, the area of this portion is $\cos(\theta)$ times the area of a half of a disk. Therefore, the probability associated with this portion is $(1/2)\cos(\theta)$. The total is just the contribution from each half and thus is $(1/2) + (1/2)\cos(\theta)$. I learned this strikingly simple explanation from John Westwater.

THE SINGLET STATE

Probably the next to simplest quantum system is the spin configurations of two spin $1/2$ particles. The importance of this system for us is that it may be used to demonstrate the most peculiar feature of quantum mechanics: there is no objective property of a system that determines the results of all experiments. This feature had been noted from the early days of the subject. The critical discussion by Einstein, Podolsky, and Rosen was particularly influential. Subsequently, Bohm proposed that the system of two spin $1/2$ particles in the singlet state was a natural system for illustrating this feature. These suggestions were part of the background for Bell's work described below.

The following is a thought experiment that bears at least some resemblance to actual experiments that have been conducted. The experimental setup involves two spin $1/2$ particles that are prepared in such a way that they have opposite spins. The particles are then separated by a great distance. The distance is supposed to be chosen so that measurements can be conducted in a time that does not allow signaling between the two locations. (One should think of this as the distance across a laboratory; the measurements are made very rapidly by some automatic device.) At the given time, each particle's spin is measured by a device involving a magnetic field gradient. The new feature is that in this experiment the magnetic field gradient can be taken in two different directions, one for each particle.

This experiment is also like a huge family of coin-toss experiments, except now there is a pair of coins corresponding to each choice of a pair of directions. The choice of the pair of directions determines the experiment and hence the coin toss. This being quantum mechanics, there is no claim that there is an overall outcome of the experiment that determines all the coin tosses corresponding to all the directions simultaneously.

We shall refer to the two quantum mechanical systems describing the particles separately as the left and right systems. The set of all states for each individual system is a two-dimensional sphere with the transition probability function given by $\cos^2(\theta/2)$. The set of all states for the combined system is a more complicated six-dimensional object with its own transition probabilities. However, for each pair of states u and v of the left and right systems, there is a corresponding state $u \otimes v$ of the combined system. Such a state of the combined state is called a *product state*. In this state the spin of the left particle in the u direction is 1, the spin of the left particle in the v direction is also 1, and the two particles act as independent units.

The construction of such product state is part of the mathematics of quantum mechanics. The transition probability between product states $u \otimes v$ and $e \otimes f$ of the combined system is

$$p(u \otimes v, e \otimes f) = p(u, e) \cdot p(v, f),$$

the product of the separate transition probabilities for the individual systems.

Consider a state e of the left system and a state f of the right system. Let e' be the state of the left system opposite to e, and let f' be the state of the right system opposite to f. Then the four states $e \otimes f$, $e \otimes f'$, $e' \otimes f$, and $e' \otimes f'$ form a maximal opposite set for the combined system.

There are also states of the combined system that are not product states; there can be a more complicated dependence. We shall be interested in a special state s called the *singlet state*. This state has the special feature that it does not prefer a particular direction in space. Each spin of an individual particle in a particular direction has a probability $1/2$ of being in one direction and $1/2$ of being in the opposite direction. The total spin of the two particles in any particular direction is zero; the spins of the individual particles in a given direction are opposite. Measurement of the spins of the two particles in nearly the same direction tend to give nearly opposite answers.

These properties have a mathematical formulation that follows from a single principle: the transition probability from the singlet state s to a product state $e \otimes f$ is

$$p(s, e \otimes f) = \frac{1}{2} p(e, f'),$$

where f' is opposite to f. Here $p(e, f')$ is the transition probability for a single spin $1/2$ particle from state e on the sphere to the state f' opposite to f on the sphere.

Here are examples. We have

$$p(s, e \otimes e') = \frac{1}{2}.$$

The singlet state has equal transition probabilities to all product states with opposite spins. It favors opposite pairs in an even-handed way. On the other hand, the transition probability

$$p(s, e \otimes e) = 0.$$

The singlet state is opposite to parallel pairs of the combined system.

For each choice of the directions e and f, we get a probability space

$$\Omega_{e,f} = \{e \otimes f, e' \otimes f, e \otimes f', e' \otimes f'\}$$

with four points. The observables of the combined system that correspond to observables for the first and for the second system are denoted σ_e^L and σ_f^R. The observable σ_e^L has the value 1 on the two points $e \otimes f$ and $e \otimes f'$ and the value -1 on the other two points. The observable σ_f^R has the value 1 on the two points $e \otimes f$ and $e' \otimes f$ and the value -1 on the other two points.

For fixed e and f, the observables σ_e^L and σ_f^R have a joint probability distribution. In the singlet state the probabilities only depend on the angle ρ between e and f. They are given by

$$\mathbf{P}_s[\sigma_e^L = 1 \textbf{ and } \sigma_f^R = 1] = p(s, e \otimes f) = \frac{1}{2}p(e, f') = \frac{1}{2}\sin^2(\rho/2).$$

Similarly,

$$\mathbf{P}_s[\sigma_e^L = -1 \textbf{ and } \sigma_f^R = -1] = p(s, e' \otimes f') = \frac{1}{2}p(e', f) = \frac{1}{2}\sin^2(\rho/2).$$

In the same way we get

$$\mathbf{P}_s[\sigma_e^L = 1 \textbf{ and } \sigma_f^R = -1] = p(s, e \otimes f') = \frac{1}{2}p(e, f) = \frac{1}{2}\cos^2(\rho/2)$$

and

$$\mathbf{P}_s[\sigma_e^L = -1 \textbf{ and } \sigma_f^R = 1] = p(s, e' \otimes f) = \frac{1}{2}p(e', f') = \frac{1}{2}\cos^2(\rho/2).$$

Some of the events for which probabilities may be computed depend only on the left observable σ_e^L. The event $\sigma_e^L = 1$ is the set $\{e \otimes f, e \otimes f'\}$, and the event $\sigma_e^L = -1$ is the set $\{e' \otimes f, e' \otimes f'\}$. We shall want to consider such events as belonging to a closed system $\mathcal{F}_L$ of events measured on the left.

Similarly, other events depend only on the right observable σ_f^R. The event $\sigma_f^R = 1$ is the set $\{e \otimes f, e' \otimes f\}$, and the event $\sigma_f^R = -1$ is the set $\{e \otimes f', e' \otimes f'\}$. These events belong to a corresponding closed system $\mathcal{F}_R$ of events measured on the right.

It is easy to use additivity to compute the probabilities of the events in these two closed systems. For each choice of sign $\pm$,

$$\mathbf{P}_s[\sigma_e^L = \pm 1] = \frac{1}{2}$$

and

$$\mathbf{P}_s[\sigma_f^R = \pm 1] = \frac{1}{2},$$

for arbitrary directions e and f. This is a reflection of the fact that the singlet state prefers no particular direction; this forces the probabilities in opposite directions to be the same.

There are events that belong to neither of these closed systems and instead describe how the right-hand and left-hand results compare to each other. The event $\sigma_e^L = \sigma_f^R$ is $\{e \otimes f, e' \otimes f'\}$. The event $\sigma_e^L = -\sigma_f^R$ is $\{e \otimes f', e' \otimes f\}$. We can use additivity to obtain

$$\mathbf{P}_s[\sigma_e^L = \sigma_f^R] = \sin^2(\rho/2)$$

and

$$\mathbf{P}_s[\sigma_e^L = -\sigma_f^R] = \cos^2(\rho/2).$$

This last equation has an important consequence when $\rho = 0$. In that case it says that $\mathbf{P}_s[\sigma_e^L = -\sigma_e^R] = 1$. Thus $\sigma_e^L = -\sigma_e^R$ is almost sure; the two spin measurements in the same direction give the opposite result. Another way of stating this is that the event $\{e \otimes e, e \otimes e'\}$, where $\sigma_e^L = 1$, is equivalent to the event $\{e \otimes e', e' \otimes e'\}$, where $\sigma_e^R = -1$. The two events are equivalent even though one is in $\mathcal{F}_L$ and the other is in $\mathcal{F}_R$.

In the next section it will be useful to have a special case of the above calculations. Let e and f be two points at an angle of 120 degrees. Then

$$\mathbf{P}_s[\sigma_e^L = 1 \textbf{ and } \sigma_f^R = 1] = \frac{3}{8}$$

and

$$\mathbf{P}_s[\sigma_e^L = 1 \textbf{ and } \sigma_f^R = -1] = \frac{1}{8},$$

and similarly for the other choice of sign. Thus

$$\mathbf{P}_s[\sigma_e^L = \sigma_f^R] = \frac{3}{4}$$

and

$$\mathbf{P}_s[\sigma_e^L = -\sigma_f^R] = \frac{1}{4}.$$

These expressions say that when the measurement directions are nearly opposite, they tend on the average to give the same answer. This is just another way to say that the spins in the singlet state have a fairly strong chance to be opposite. For each possible measurement with these relative angles, this looks like dependent coin tossing. The remaining question is how these possible measurements relate to each other. This question is an entry into the deeper quantum mysteries.

BELL'S FIRST THEOREM

There were a number of profound investigations of the peculiar structure of quantum mechanics before Bell's work, but he gave a particularly simple and striking formulation. Part of the charm of his results was that they suggested experiments that could actually be performed. Most of Bell's papers on this subject have been collected in one volume [6], and this is a convenient reference for his work described in the following.

First Bell proved a theorem that implies that it is impossible to produce a probability interpretation in terms of a single probability space that agrees with quantum mechanics. We present a version of his result.

Theorem: *Assume that there is a set Ω of outcomes and a probability measure defined on subsets of this set (events). Assume that there are events A equivalent to A', B equivalent to B', and C equivalent to C'. Then*

$$\mathbf{P}[A \textbf{ and } \textbf{not } B'] + \mathbf{P}[B \textbf{ and } \textbf{not } C'] + \mathbf{P}[C \textbf{ and } \textbf{not } A'] \leq 1.$$

Proof: Since in computing probabilities one can substitute equivalent events, it is sufficient to show that

$$\mathbf{P}[A \textbf{ and } \textbf{not } B] + \mathbf{P}[B \textbf{ and } \textbf{not } C] + \mathbf{P}[C \textbf{ and } \textbf{not } A] \leq 1.$$

However, the three events whose probabilities are being computed are exclusive, so the sum is equal to the probability of the union event. This event has probability less than 1. **End proof.**

This statement and proof of Bell's first theorem are different but equivalent to his original statement. They make clear that this theorem is just a very special result of elementary probability theory and applies for perfectly general probability experiments. Why then is it of interest? The reason is that it tells something about quantum mechanics.

Corollary: *Consider the quantum mechanical system describing the spin observables σ_e^L and σ_f^R of two spin $1/2$ particles on the left and right, for all pairs of directions e and f. Assume that this system is in the singlet state s. Then there is no probability measure* **P** *defined on a closed system $\mathcal{F}$ of subsets of a set Ω that gives the probabilities predicted by quantum mechanics.*

Proof: Let e, f, and g be three directions in a plane at angles of 120 degrees to each other. Let A, B, and C be the events that $\sigma_e^L = 1$, $\sigma_f^L = 1$, and $\sigma_g^L = 1$. The equivalent events A', B' and C' are $\sigma_e^R = -1$, $\sigma_f^R = -1$, and $\sigma_g^R = -1$. We did the quantum mechanical calculation at the end of the previous section; each of the three probabilities in the statement of the theorem is $3/8$. The sum of these probabilities is $9/8$. However, if there were a probability model, then according to Bell's inequality the sum could not exceed 1. Thus there cannot be a single underlying probability model. **End proof.**

The logic of the argument is simple. The two particles have been well separated. The spin of each particle can be measured by passing it through an appropriate magnetic field gradient. There are three possible pairs of settings of the magnets, corresponding to directions e on the left and f on the right, f on the left and g on the right, and g on the left and e on the right. According to quantum mechanics, for each pair of settings the spin in the direction on the left and the spin in the direction on the right are dependent to a certain degree. If there were some intrinsic property of the particles that determined the result for each conceivable pair of settings, then there would be combined events defined in terms of the three settings. However, Bell's first theorem says that the dependence is strong enough so that it is incompatible with the existence of such combined events. So there is no such intrinsic property.

If we think of the spin experiment associated with each pair of directions as a dependent coin-toss experiment, this says that there is no one overall result of the experiment that simultaneously determines the values of all the coin tosses corresponding to all the pairs of directions.

Bell's first theorem implies that the experiment with two spin $1/2$ particles cannot be explained by a probability model, but the fact that quantum mechanics cannot be reduced to probability on a single probability space is far more general. In fact, the only exception is the example of the single spin $1/2$ particle.

In a reasonable conception of a real world, experiments should have outcomes. Does quantum mechanics thus imply that there is no real world, in the sense that there is no one outcome intrinsic to the particular system that determines the values of all possible measurements? This seems to be the case in the usual formulations of quantum mechanics, and Bell's first theorem and similar results seem to rule out any alternative view. However, there is a huge gap in the argument up to this point. As Bell pointed out, quantum mechanics makes an implicit assumption that the probabilities that emerge from measurement are intrinsic to the system being measured. Thus the probabilities for a pair of spin $1/2$ particles in the singlet state are supposed to be properties of that system. However, this assumption neglects the fact that the act of performing the measurement may itself be part of the mechanism that generates the probabilities. The measured results for the pair of spin $1/2$ particles may depend on the way that the magnets that do the measuring are arranged. So nothing in Bell's first theorem or in any similar result rules out the possibility of a family of probability models that reproduce the results of quantum mechanics.

ACTIVE LOCALITY

The possibility that the probabilities depend on the way the measurement is conducted requires an analysis involving more than one probability measure. This is the framework for discussing active locality. (We continue to use Nelson's terminology for the different kinds of locality, which is a variant of Bell's.)

We shall assume that we are dealing with a set Ω of outcomes and a closed system $\mathcal{F}$ of events, each of which is a subset of Ω. All the events specified in the following belong to $\mathcal{F}$. The experiment takes place in two stages. The first stage is the preparation stage. We shall refer to this closed system of *preparation events* as $\tilde{\mathcal{F}}$. The subsequent second stage is the measurement stage. Measurements are made simultaneously at two distant locations, on the left and on the right. Some of the events at this stage depend only on what happens at the left; others depend only on what happens at the right. We shall refer to the closed system of *left events* as $\mathcal{F}_L$ and the closed system of *right events* as $\mathcal{F}_R$.

The new feature is that there may be several probability measures, corresponding to the various measurements that are made on the left and on the right. If the measurement on the left is a measurement of type a and the measurement on the right is a measurement of type b, then the probability of an event A is written $\mathbf{P}_{ab}[A]$.

The assumption of *active locality* says that for every event A defined in terms of events that are either in $\tilde{\mathcal{F}}$ or in $\mathcal{F}_L$, the probability $\mathbf{P}_{ab}[A]$

does not depend on b, and for every event B defined in terms of events that are either in $\tilde{\mathcal{F}}$ or in $\mathcal{F}_R$, the probability $\mathbf{P}_{ab}[B]$ does not depend on a. If active locality were violated, an intervention at a distant location could influence an event at the same or earlier time.

Example: Coin tossing This is the most elaborate version yet of the coin-tossing example. As before, an outcome is a report of the results of a coin toss in Tucson at the preparation stage together with reports of the results of coin tosses in Seattle and New York at the measurement stage. The events defined by the results in Tucson form $\tilde{\mathcal{F}}$, those defined by the results in Seattle form $\mathcal{F}_L$, and those defined by the results in New York form $\mathcal{F}_R$.

A fair coin is flipped in Tucson at the preparation stage. If the coin comes up heads then identical coins biased toward heads are shipped to Seattle and New York. If it comes up tails, then identical coins biased toward tails are shipped to Seattle and New York.

Now in Seattle and also in New York a decision is made whether to use the biased coin that was shipped from Tucson or to use a fair coin. Call the decision to use the biased coin the 1 decision and the decision to use the fair coin the 0 decision. Then the probability measure $\mathbf{P}_{00}$ describes the situation when the fair coin is used in both locations, the probability measure $\mathbf{P}_{10}$ describes when the biased coin is used only in Seattle, the probability measure $\mathbf{P}_{01}$ describes when the biased coin is used only in New York, and the probability measure $\mathbf{P}_{11}$ describes when the biased coin is used in both locations.

It is easy to see that the probability measures $\mathbf{P}_{00}$, $\mathbf{P}_{10}$, and $\mathbf{P}_{01}$ make the events in $\mathcal{F}_L$ measured in Seattle independent of the events in $\mathcal{F}_R$ measured in New York. These measures describe a situation in which one of the locations has refused to participate in the conspiracy. The probability measure $\mathbf{P}_{11}$, however, describes a situation in which both parties used biased coins that have been prepared in the same way. This is dependent coin tossing.

According to the condition of active locality, the choice of one party of whether to join in the conspiracy should not change the observations of the second party that are made at the same time. This is indeed the case;

the $\mathbf{P}_{10}$ probability of a head in Seattle is $1/2$ and the $\mathbf{P}_{11}$ probability of a head in Seattle is still $1/2$, independent of how the experiment is conducted in New York. In one case there is no dependence, while in the other case there is a dependence. The dependencies that might be uncovered by a later analysis of the results in the two cities are seen in events that belong to $\mathcal{F}$ but not to $\mathcal{F}_L$ or $\mathcal{F}_R$ individually.

Active locality is related to the impossibility of signaling or intervening instantaneously over long distances. An action in one location can turn on or switch off a dependence, but it cannot be used to signal.

BELL'S SECOND THEOREM

We have seen that the first Bell theorem rules out the possibility that quantum mechanics is described by a unique probability measure. Could quantum mechanics instead be described by a family of probability measures depending on the measurement? Indeed this is possible; there are theories such as Bohm's theory [7], [8] and stochastic mechanics [9], [10] in which the atomic particles have a probability description in terms of random particle trajectories. These theories give the same results for fixed-time position measurements as quantum mechanics. Such theories can even give the correct results for spin measurements [11], [12], [13] by incorporating a suitable spin trajectory.

There is no reason to believe that in such a theory the measured spin must be an intrinsic property of the particle that agrees with all the predictions of quantum mechanics. The spin is measured by inserting a large apparatus involving a magnetic field, and the presence of this apparatus could influence the probabilities of various outcomes.

It might be possible, for instance, that the spins of two spin $1/2$ particles in the singlet state are actually continuous random variables that are constrained to be opposite, at least up to the time of the measurement. As each particle passes through the measuring apparatus involving the magnetic field gradient, its spin component in the direction of the gradient takes on one or the other discrete value and the particle is deflected accordingly.

This picture denies the existence of simultaneous coin-toss experiments corresponding to all possible directions. Two magnetic field gradient directions are chosen, one for each particle, and corresponding to these are the two coin tosses.

Such theories are often regarded as problematic. In fact, Bell formulated a second theorem that shows that, even if the probabilities do depend on the measurement apparatus, it is still impossible to produce a probability model for quantum mechanics unless either the active or passive locality assumption is violated.

Active and passive locality The simplest experimental context is again the system of two particles in the singlet spin state. They are prepared at some initial time and then allowed to travel to two widely separated regions, say on the left and on the right. At a later time, when they are separated, decisions are made about which component of spin to measure, say a on the left and b on the right. As usual we assume that we are dealing with a set Ω of outcomes and a closed system $\mathcal{F}$ of events, each of which is a subset of Ω. All the events specified in the following belong to $\mathcal{F}$. The probability of an event A may depend on a and b, and so it is written $\mathbf{P}_{ab}[A]$.

We briefly review the now-standard notation for the various closed systems of events. Events at the preparation stage take place at the time of the preparation of the two particles in the singlet state or after that time, but before the spin measurement is conducted. This closed system of *preparation events* is $\tilde{\mathcal{F}}$. Some events at the subsequent measurement stage depend only on the particle at the left; others depend only on the particle at the right. The closed system of *left events* is $\mathcal{F}_L$, and the closed system of *right events* is $\mathcal{F}_R$.

Now we recall the two locality assumptions. The first assumption is *active locality*. It says that for every event A defined in terms of events that are either in $\tilde{\mathcal{F}}$ or in $\mathcal{F}_L$, the probability $\mathbf{P}_{ab}[A]$ does not depend on b, and for every event B' defined in terms of events that are either in $\tilde{\mathcal{F}}$ or in $\mathcal{F}_R$, the probability $\mathbf{P}_{ab}[B']$ does not depend on a.

In the spin experiment involving widely separated particles in the singlet state, a violation of active locality might mean that the choice of direction b of the magnetic field gradient used to measure the spin of the right-hand particle would instantly change the probabilities of events A associated with the left-hand particle. There would be no doubt about the direction b of the magnetic field gradient. If also the events A were events that could be observed in a practical way, then this would have a great impact on technology; one could send signals faster than the speed of light.

The second assumption is *deterministic passive locality* (which is a consequence of the more general condition of passive locality). Let A be an event in $\mathcal{F}_L$ and A' be an event in $\mathcal{F}_R$. Assume that A and A' are equivalent with respect to $\mathbf{P}_{aa}$. Then there is an event $\tilde{A}$ in $\tilde{\mathcal{F}}$ such that A and A' are each equivalent to $\tilde{A}$ with respect to $\mathbf{P}_{aa}$. If deterministic passive locality were violated, it would be possible that there would be randomness at the measurement stage with no origin in any preparation stage, yet which would still maintain the perfect correspondence between events at the distant locations.

Now we are able to prove Bell's second theorem.

Theorem: *Let $\mathbf{P}_{ab}$ be a family of probability measures on the same closed system $\mathcal{F}$ corresponding to measurement parameters a and b. Let $\tilde{\mathcal{F}}$ be a closed system of prior preparation events. Let $\mathcal{F}_L$ and $\mathcal{F}_R$ be closed systems of subsequent events (left and right). Assume that the probability measures $\mathbf{P}_{ab}$ satisfy active locality and the probability measures $\mathbf{P}_{aa}$ satisfy deterministic passive locality. Let A, B, and C be events in $\mathcal{F}_L$, and A', B', and C' be events in $\mathcal{F}_R$ such that the pairs A, A', B, B', and C, C' are equivalent with respect to $\mathbf{P}_{aa}$, $\mathbf{P}_{bb}$, and $\mathbf{P}_{cc}$. Then*

$$\mathbf{P}_{ab}[A \textbf{ and } \textbf{not } B'] + \mathbf{P}_{bc}[B \textbf{ and } \textbf{not } C'] + \mathbf{P}_{ca}[C \textbf{ and } \textbf{not } A'] \leq 1.$$

The idea of the proof is to show that the equivalence of the simultaneous separated events implies that the randomness must have all originated at the preparation stage. Thus the introduction of the measuring devices cannot have influenced the probabilities.

Proof: Active locality implies that every event $\tilde{A}$ that belongs to the preparation stage $\tilde{\mathcal{F}}$ has a probability $\mathbf{P}_{ab}[\tilde{A}] = \mathbf{P}[\tilde{A}]$ that is independent of the decisions about measurement made at a later stage. The strategy is to show that all the relevant events are equivalent to events in $\tilde{\mathcal{F}}$.

We analyze the first term in the sum of probabilities. The other two follow the same pattern.

Let A in $\mathcal{F}_L$ be equivalent to A' in $\mathcal{F}_R$. By deterministic passive locality there is an event $\tilde{A}$ in $\tilde{\mathcal{F}}$ that is $\mathbf{P}_{aa}$ equivalent to A. This means that the events A, $\tilde{A}$, and A **and** $\tilde{A}$ all have the same $\mathbf{P}_{aa}$ probability.

The events A, $\tilde{A}$, and A **and** $\tilde{A}$ are each defined in terms of events in the closed systems $\mathcal{F}_L$ or $\tilde{\mathcal{F}}$. By active locality these events have the same $\mathbf{P}_{ab}$ probabilities as $\mathbf{P}_{aa}$ probabilities. Thus they all have the same $\mathbf{P}_{ab}$ probabilities. This means that $\tilde{A}$ is also $\mathbf{P}_{ab}$ equivalent to A.

The same type of argument shows that if B' is in $\mathcal{F}_R$, then there is an event $\tilde{B}'$ in $\tilde{\mathcal{F}}$ that is $\mathbf{P}_{bb}$ and hence $\mathbf{P}_{ab}$ equivalent to B.

We have shown that the event A is $\mathbf{P}_{ab}$ equivalent to the event $\tilde{A}$ and that the event B' is $\mathbf{P}_{ab}$ equivalent to the event $\tilde{B}'$. It follows from elementary probability theory that the event A **and not** B' is $\mathbf{P}_{ab}$ equivalent to the event $\tilde{A}$ **and not** $\tilde{B}'$. This last event is in $\tilde{\mathcal{F}}$. Furthermore, the probabilities of equivalent events are equal. Thus

$$\mathbf{P}_{ab}[A \textbf{ and } \textbf{ not } B'] = \mathbf{P}[\tilde{A} \textbf{ and } \textbf{ not } \tilde{B}'].$$

As we have already observed, active locality implies that the last probability does not depend on a or b.

This works with each of the three terms. All three probabilities may be computed with the probability space Ω with events $\tilde{\mathcal{F}}$ and probability measure $\mathbf{P}$. Bell's second theorem reduces to Bell's first theorem. **End proof.**

Corollary: *Consider the quantum mechanical system describing the spin observables σ_e^L and σ_f^R of two spin $1/2$ particles on the left and right, for all pairs of directions e and f. Assume that this system is in the singlet state s. Then there is no family of probability measures $\mathbf{P}_{ef}$ depending on* *the measurement directions defined on a closed system $\mathcal{F}$ of subsets of a set Ω that satisfy active and deterministic passive locality and give the probabilities predicted by quantum mechanics.*

Proof: In the application to the two separated spin $1/2$ particle experiment, we take e, f, and g to be three directions at 120 degree angles to each other. The events A, B, and C are $\sigma_e^L = 1$, $\sigma_f^L = 1$, and $\sigma_g^L = 1$, and the events A', B', and C' are $\sigma_e^R = -1$, $\sigma_f^R = -1$, and $\sigma_g^R = -1$. Suppose there were three probability measures $\mathbf{P}_{ab}$, $\mathbf{P}_{bc}$, and $\mathbf{P}_{ca}$ corresponding to three pairs ef, fg, and ge of measurement directions. According to quantum mechanics, the three probabilities in the statement of the theorem sum to $9/8$. This contradicts the inequality in the theorem. So we conclude that no family of local probability models can explain the experiment, even if the probabilities are allowed to depend on which measurement is made. **End proof.**

The logic of this proof is elementary but subtle. Two particles are prepared in a special way so that they are described by the quantum mechanical singlet state. They are then separated by a great distance so that no further communication is possible. Their spins measured in various directions are random but have a certain amount of dependence. The measurement of the spin of one of the particles in a particular direction is done by introducing a magnetic field with its gradient in that direction. It is possible that introducing this magnetic field affects the probabilities for the spin of this particle, but the active locality assumption says that it should not affect the probabilities for the other particle.

The first part of the argument is the observation that when the two magnetic field gradients are taken in opposite directions the results exactly coincide. The deterministic passive locality assumption says that this implies that the randomness must have been introduced at the time the particles were prepared, since a later source of the randomness would influence the two particles separately and spoil the perfect coincidence. Thus with the magnetic field gradients in opposite directions, each spin result is determined by something that happened at the preparation state.

The second part of the argument uses the active locality assumption. The spins may also be measured in the situation when the magnetic field gradient for one particle is not in the same or opposite direction to the magnetic field gradient for the other particle. According to active locality the results of spin measurements for one particle should not depend on the direction used for the measurement on the other particle. Therefore, for the magnetic field gradients in arbitrary directions the spin results for each particle are still determined by something that happened at the preparation stage. So the magnets are not responsible for the randomness; if there is an overall outcome of the experiment responsible for the randomness it must be intrinsic to the system of two particles.

In the case of magnetic field gradients that are not the same or opposite the dependence is strong but not perfect. This dependence is entirely due to something that happens at the preparation stage; the magnet configurations do not affect the probabilities. In this situation Bell's first theorem says that the dependence is strong enough so that there cannot be an intrinsic overall outcome that determines the result for each conceivable measurement. This is the last step in proving Bell's second theorem.

Technical note:

Recently Greenberger, Horne, Shimony, and Zeilinger [14] gave examples of quantum events that occur with probability 1 to which arguments like Bell's could be applied. To do this they need a more intricate experiment with three particles. The active and passive locality assumptions and the general conclusions about quantum mechanics are the same [15].

Quantum mechanics and the real world It follows from this argument that probabilistic theories that agree with quantum mechanics for observable quantities must be nonlocal. An analysis of the probabilistic theories involving a particle trajectory or spin trajectory would show that they violate passive locality for the observed quantities, such as fixed time position measurements. Worse, they also violate active locality for the particle trajectories and spin trajectories. These trajectories are not observed, since they involve measuring the position or spin of the particle at successive times, but they are supposed to be part of the reality that these theories describe. The violation of active locality makes these theories difficult to accept.

Again: Does quantum mechanics imply that there is no real world, in the sense that there is no intrinsic overall outcome of an experiment on a system? This seems to be the case if one accepts quantum mechanics as the ultimate truth. Bell's second theorem shows that the search for alternatives in the framework of probability theory is difficult, even if one recognizes that in such an alternative theory the probability measure could be influenced by the choice of measuring apparatus. Any such alternative theory must be nonlocal, violating either active or passive locality.

Bell's theorem should not be regarded as saying that since alternative theories violate active or passive locality they must be rejected in favor of quantum mechanics. Quantum mechanics itself appears to violate passive locality in some sense. It is difficult to make this precise because of the other conceptual difficulties of quantum mechanics. However, consider the experiment in which spin measurements on two particles are made with the measured component of spin being taken along some fixed direction. This experiment has a probability description, even in the framework of quantum mechanics. The spins are random, yet the two particles always have opposite spins. This perfect match seems to have no common cause in past events, so passive locality is violated.

It seems plausible that a reasonable alternative theory should respect active locality. On the other hand, an alternative theory could well violate passive locality. After all, this would be less disturbing than the picture given by quantum mechanics, which implies a denial of the real world. Why is this not the way out? The problem is that quantum mechanics, however mysterious its foundation, works very well to describe the results of experiments. There is no well-developed alternative theory on the scene. Meanwhile, quantum mechanics does its job.

Could such a theory really be believed by the overwhelming majority of physicists? Clearly, yes. The theory makes predictions, often of astonishing accuracy, in many domains. It is also quite robust, in that many of the predictions do not seem to depend on the details of how the experiments are performed. Quantum mechanics must be counted as a spectacular success.

It is sometimes claimed that all that one could ask of a theory such as quantum mechanics is that it predict probabilities involving specified measurements, say pointer readings or perhaps cloud chamber tracks. However, in everyday work most physicists do act as if they believe in an objective atomic world. They do not seem to need a sharp separation between an atomic level world described by quantum mechanics and a macroscopic world of measuring devices. They use quantum mechanics to explain properties of bulk matter, and they also use it to explain direct quantum mechanical effects on a fairly large scale (perhaps comparable to the size of laboratory apparatus).

A physicist or a chemist believes that the electrons of sodium bond with the electrons of chlorine to form a salt molecule. The "wave function" that expresses the quantum mechanical state of the electrons somehow determines the strength of the bonding. Ordinary table salt is made of many such molecules, and there is no question of measurement or any other quantum mystery. Huge salt flats on the surface of the Earth existed long before there were physicists and measuring instruments and research grants. Even if it is essential to specify experimental conditions for some purposes, there must be a world apart from measurement. The trouble is that quantum mechanics does not give a satisfactory picture of how this comes about.

CONCLUSION

Bell's second theorem says that if there is a probability model satisfying the locality properties, then it must violate a result of quantum mechanics. What conclusion can we draw?

One possibility is that the quantum mechanical prediction is wrong for such experiments. This could eventually be settled empirically; most experiments to date do not make this seem to be a likely outcome.

Another possibility is that quantum mechanics is a correct description of Nature; there is no notion of outcome independent of the measurement of observables. This leads to an acute problem of how to describe measurement.

Still another possibility is a theory with a well-defined notion of outcome, but that violates locality. A violation of active locality would be very upsetting, since it would mean that active intervention at one point could influence the outcome at distant points. However, a violation of passive locality would only mean that dependence between simultaneous events at distant locations need have no explanation in terms of prior events. This is not clearly ruled out, but it is not evident how to construct such a theory.

The last possibility, and clearly the correct one, is that there is some totally new way of looking at Nature. It will be just as subtle as quantum mechanics, perhaps violate locality in some special way, but the resulting world view will be sharp and clear. School children will assimilate it, and fundamental science will have reached its end.

ACKNOWLEDGEMENTS

I am grateful to Hank Atha for many insightful comments and to Larry Grove for useful mathematical advice. Also, I thank Richard Healey and John Westwater for setting me right on various points.

REFERENCES

[1] Edward Nelson, "Field theory and the future of stochastic mechanics," pp. 438–469, in: *Stochastic Processes in Classical and Quantum Systems, Proc., Ascona, Switzerland, 1985*, S. Albeverio, G. Casati, and D. Merlini, eds., Lecture Notes in Physics #262, Springer, Berlin, 1986.

[2] Edward Nelson, "Stochastic mechanics and random fields," pp. 427–450 in *École d'Été de Probabilités de Saint-Flour XV–XVII, 1985–87*, Lecture Notes in Mathematics #1362, Springer, Berlin, 1988.

[3] Hans Reichenbach, *The Direction of Time*, University of California Press, Berkeley, 1971.

[4] Michael Redhead, *Incompleteness, Nonlocality, and Realism*, Clarendon Press, Oxford, 1987.

[5] S. Kochen and E. P. Specker, "The problem of hidden variables in quantum mechanics," *Journal of Mathematics and Mechanics* **17** (1967), 59–87.

[6] J. S. Bell, *Speakable and Unspeakable in Quantum Mechanics*, Cambridge University Press, Cambridge, 1987.

[7] Peter R. Holland, *The Quantum Theory of Motion: An Account of the de Broglie-Bohm Causal Interpretation of Quantum Mechanics*, Cambridge University Press, Cambridge, 1993.

[8] D. Bohm and B. J. Hiley, *The Undivided Universe: An Ontological Interpretation of Quantum Theory*, Routledge, London, 1993.

[9] Edward Nelson, *Dynamical Theories of Brownian Motion*, Princeton University Press, Princeton, NJ, 1967.

[10] Edward Nelson, *Quantum Fluctuations*, Princeton University Press, Princeton, NJ, 1985.

[11] T. G. Dankel, "Mechanics on manifolds and the incorporation of spin into Nelson's stochastic mechanics," *Archive for Rational Mechanics and Analysis* **37** (1970), 192–222.

[12] William G. Faris, "Spin correlation in stochastic mechanics," *Foundations of Physics* **12** (1982), 1–26.

[13] William G. Faris, "A stochastic picture of spin," pp. 154–168, in: *Stochastic Processes in Quantum Theory and Statistical Physics*, S. Albeverio, Ph. Combe, and M. Sirugue-Collin, eds., Springer, Berlin, 1982.

[14] Daniel M. Greenberger, Michael A. Horne, Abner Shimony, and Anton Zeilinger, "Bell's theorem without inequalities," *American Journal of Physics* **58** (1990), 1131–1143.

[15] Robert K. Clifton, Michael L. G. Redhead, and Jeremy N. Butterfield, "Generalization of the Greenberger-Horne-Zeilinger algebraic proof of nonlocality," *Foundations of Physics* **21** (1991), 149–184.

Notes

Introduction

[p. xiv] The little book of Einstein's articles: see Lorentz et al. (1923), listed in the Bibliography.

[p. xiv] The monograph I read in graduate school: Werner Heisenberg's *The Phuysical Principles of the Quantum Theory* (1929).

[p. xvi] Einstein anecdote: see Pais, *Rev. Mod. Phys.* **51** 863 (1979).

Chapter 1 Prologue I: Atoms

[p. 1] The first human to catch a direct glimpse of a single, isolated atom may have been Dr. Werner Neuhauser, in late 1979 at the University of Heidelberg in Germany. See Chapter 15, "The Impossible Observed."

[p. 3] Concerning Boltzmann's equation, nearly a century passed before the qualifying phrase was fully clarified by Harold Grad and Oscar Lanford III in the 1960s and 70s. They resolved the paradox of how a system whose dynamics works equally well "in reverse"—that is, if all velocities are suddenly reversed—can establish an "arrow of time." The solution is that, although the velocities can in principle be reversed at any later time, the operation cannot yield a "properly chaotic" condition; for such an untypical state of a gas, the entropy must clearly decrease at later times!

[p. 3] Useful works on Boltzmann and Mach include: Engelbert Broda's *Ludwig Boltzmann: Man, Physicist, Philosopher* (1983); John Blackmore's *Mach: His Work, Life and Influence* (1972); and Cohen and Seeger (eds.), *Ernst Mach: Physicist and Philosopher* (1970).

[p. 4] Kac's riposte to Feynman: see the review of Kac's memoir *Enigmas of Chance* (1985) in *Ann. Prob.* **14**, no. 4 (1986), p. 1147.

Chapter 2 Prologue II: Quanta

The histories of Abraham Pais provide a wealth of detail on the events described in this chapter: *"Subtle is the Lord . . . ": The Science and the Life of Albert Einstein* (1982); *Inward Bound of Matter and Forces in the Physical World* (1986);

and *Niels Bohr's Times, In Physics, Philosophy and Polity* (1991). Another useful reference is D. ter Haar, *The Old Quantum Theory*, which reproduces many original papers, including Rutherford's on nuclear scattering. For quantum theory up to 1912, see Kuhn (1978).

[p. 6] " A senior scientist judged the result": This statement has been attributed to the Astronomer Royal of England, to Lord Kelvin, and to others. It may be apocryphal, but a mood of general satisfaction at the time would not be surprising.

[p. 9] Thomas Kuhn has argued that it was Einstein, not Planck, who took the revolutionary step of introducing quanta into physics; see Kuhn (1978).

[p. 9] In 1922, the Nobel Committee, ever disdainful of theory, cited Einstein for explaining the photoelectric effect and "for his services to theoretical physics," which they did not further describe. Politics and prejudice may also have delayed the award; see Pais (1982).

[p. 10] Rutherford quotation: Andrade (1964), p. 111.

[p. 12] Stern and von Laue respond to Bohr: Pais (1986), p. 208. Bohr anecdotes: Pais (1986), pp. 208–10.

[p. 12] Einstein's lack of pluck: Pais (1986), p. 208.

[p. 13] Experiments such as that of Stern and Gerlach that revealed the peculiar "discreteness" of atoms are described in any standard quantum mechanics text, for instance Albert Messiah's *Quantum Mechanics*, Vols. I and II.

Chapter 3 Revolution, Part I: Heisenberg's Matrices

Biographic information can be found in Armin Hermann's *Werner Heisenberg 1901–1976* (1976); Elisabeth Heisenberg's *Inner Exile: Recollections of a Life with Werner Heisenberg* (1980); and, more recently, David Cassidy's *Uncertainty: The Life and Times of Werner Heisenberg* (1992).

[p. 18] Bohr quote: in S. Rozental, ed. (1967), p. 95. The thesis-defense reminiscence is from an interview by Thomas Kuhn (Sources for the History of Quantum Physics, Interview 3.)

[p. 19] The phrase "quantum mechanics" appears for the first time in print in a paper of Born's in 1924, which includes a footnote thanking his assistant Heisenberg for help on the calculations, see van der Waerden (1967).

[p. 20] Heisenberg's remarkable first paper appeared in *Z. für Phys.* **33** (1925), 879; it is reprinted and translated into English together with the other early papers mentioned in the text in B. L. van der Waerden (ed.), *Sources of Quantum Mechanics* (1967). Heisenberg's memoirs and other writings contain many insights into his character and views; in particular see *Physics and Philosophy: The Revolution in Modern Science* (1962); *The Physicist's Conception of Nature* (1970); and *Physics and Beyond: Encounters and Conversations* (1971).

[p. 23] More on the classical nature of quantum equations appears in Chapter 18, "Principles."

Chapter 4 Revolution, Part II: Schrödinger's Waves

The basic work is Walter Moore's superb biography *Schrödinger, Life and Thought* (1989). Abraham Pais' and Max Jammer's histories also contain relevant information. Schrödinger's four papers appeared in *Annalen der Physik*, 1926, and are reprinted and translated into English together with de Broglie's paper in Gunther Ludwig's *Wave Mechanics* (1968), from which the quotations are taken.

[p. 30] Nasty remarks about Schrödinger's physics: "wretched," Born to Einstein, November 1926, Pais (1986) p. 260; "abomination," Heisenberg to Pauli, June 1926, Moore (1986), p. 221.

[p. 31] The equation Schrödinger rejected: see his fourth paper, Ludwig (1968), p. 153, footnote. It is linear, but contains fourth derivatives with respect to the space variable. He found it in Courant and Hilbert (1923), the "bible" of the mathematical physicist in those days.

[p. 31] Schrödinger's equation for a single particle is

$$\frac{h\sqrt{-1}}{2\pi}\frac{\partial\Psi}{\partial t} = \left(-\frac{h^2}{8\pi^2 m}\sum_{i=1}^{3}\frac{\partial^2}{\partial x_i^2} + V(\vec{x})\right)\Psi,$$

where Ψ is a complex function of $\vec{x}$, a point in space, and the time t. $V(\vec{x})$ is the potential energy of the particle at $\vec{x}$. For further discussion of the mathematics of this equation, see the Notes for Chapter 18, "Principles."

[p. 32] The mechanical analogy shown in Figure 2 is obtained by first discretizing space (the Laplacian being replaced by a second-order difference operator) and then by separating real and imaginary parts, a useful mathematical activity strangely absent from most textbooks (where its presence might help remove some of the mystery from Schrödinger's equation). The vertical springs and horizontal elastic bands generate linear restoring forces. The "repulsive connectors" between next-nearest neighbors generate repulsive forces growing linearly with separation. The respective spring or elastic force constants are given by

$$\frac{h^2}{64\pi^2 m\epsilon^4}, \qquad \frac{1}{4}\times \text{same}, \qquad \frac{mV^2}{h}(\vec{x})$$

for the elastic bands, the "repulsive connectors," and the vertical springs, respectively. (Here m = electron mass and ϵ = spacing of the balls.) The repulsive connectors might seem difficult to build into a mechanical model (implying, in technical terms, a "negative bulk modulus" of the component), but such atypical materials have in fact been invented; the one in the figure (in which vertical springs tilt to the horizontal as the connector expands, exerting horizontal pressure increasing linearly with length over a small range of lengths) seems the simplest realization.

[p. 32] Einstein may have had a picture like Figure 2 in mind when he disparaged Schrödinger's equation to Born in a letter of 1926: "Waves in $3n$-dimensional space with velocity regulated by potential energy (e.g., rubber bands)..."

[p. 34] Born quotation (on Schrödinger's wave): see Pais (1986), p. 256.

[p. 34] One can understand something of the amazing correspondence between Heisenberg's and Schrödinger's theories by noting the calculus formula

$$Q\left\{\frac{h}{2\pi\sqrt{-1}}\frac{\partial}{\partial Q}\right\} - \left\{\frac{h}{2\pi\sqrt{-1}}\frac{\partial}{\partial Q}\right\}Q = \frac{h\sqrt{-1}}{2\pi},$$

which holds if both sides are applied to a smooth function of the real variable Q. Making the admittedly mysterious identification

$$P \leftrightarrow \left\{\frac{h}{2\pi\sqrt{-1}}\frac{\partial}{\partial Q}\right\}$$

this formula becomes the equation of Born and Jordan. If this calculus fact leaves the meaning of "Q" or "P" as murky as before, the next two chapters, which describe how Heisenberg and Bohr struggled with the problem without total success, should provide some solace.

Chapter 5 Uncertainty

[p. 37] The letter to Pauli is quoted in Hermann's biography (1976), p. 33.

[p. 38] The uncertainty paper is translated into English and reproduced in Wheeler and Zurek (1983). Heisenberg's recollections of the weeks alone in Copenhagen are in Price and Chissick, eds. (1977).

For Heisenberg's mature views, and more discussion of the uncertainty relations, see Chapter 16.

Chapter 6 Complementarity

[p. 43] Jammer (1976) contains a detailed treatment of Heisenberg's γ-ray microscope.

[p. 45] See Chapters 16 and 19.

[p. 46] The closest Bohr came to a definitive exposition of his views (it has been argued that he never produced one) came in 1948–49, in an article in *Dialectica* (1948) and in his review of the decades-long debate with Einstein in the Einstein Festschrift volume (*Albert Einstein: Philosopher-Scientist*, P. A. Schilpp, ed., 1949), reproduced in Wheeler and Zurek. Einstein's "sharp formulation" complaint was registered in a Reply to Critics in the latter volume. The Bohr-Schrödinger exchange is quoted in Moore (1989).

Chapter 7 The Debate Begins

[p. 52] Unfortunately, Einstein never wrote a review of the debates with Bohr. When he summarized his epistemological views in 1936 (see "EPR"), he mentioned Bohr only in passing—so inevitably these events are primarily colored by Bohr's detailed recollections, which appeared in the volume dedicated to

Einstein's 70th birthday in 1949 (*Albert Einstein: Philosopher-Scientist*, P. A. Schilpp, ed., reprinted in Wheeler and Zurek). See also Rosenfeld's memoir in The Fourteenth Solvay Proceedings (1968), from which the quote is taken, and Heisenberg's (1967). The Fifth Solvay Conference proceedings are in "Electrons et Photons" (Gauthier-Villars, Paris, 1928).

[p. 52] Ernst Solvay, a Belgian industrialist, had sponsored a series of international gatherings of prominent scientists. The Fifth Solvay Conference proceedings are in "Electrons et Photons" (Gauthier-Villars, Paris, 1928).

[p.53] Einstein's absorption in quantum problems: letter to Otto Stern, quoted in Pais (1989), p. 9.

[p. 53] Wolfgang Pauli claimed to have refuted the pilot wave in a special case; his argument would be answered by Bohm 25 years later; see Chapter 10, "The Post-War Heresies."

[p. 54] See "ghostly action at a distance" in Chapter 16.

[p. 58] The other tests Einstein proposed in 1916 involved the magnitude of the deflection of light by the sun, observed three years later by Eddington, and the perihelion precession of the orbit of Mercury by 42 seconds of arc per century, already known to astronomers.

[p. 59] "A good joke should not be repeated....": Frank (1947), p. 216.

[p. 59] More comprehensive discussion of the *Gedanken* experiments, including reactions and criticisms by others, and later developments, can be found in Jammer (1966) and (1974).

[p. 59] Einstein's Nobel recommendation: see Pais (1982), p. 515.

Chapter 8 The Impossibility Theorem

I profited from Steven J. Heims' *John von Neumann and Norbert Wiener: From Mathematics to the Technologies of Life and Death* (1980), Norman Macrae's *John von Neumann* (1992), and Wigner's various memoirs.

[p. 62] Hilbert space had its origin in decomposing functions into sines and cosines with various frequencies; for example:

$$f(t) = \sin(t) + 3\cos(t) - 2\sin(2t) + \sqrt{2}\cos(2t) + \cdots,$$

where the ellipsis stands for an infinite series of omitted terms with successively higher frequencies of oscillation. The "space" of all these functions has, in a completely natural sense, an infinite set of perpendicular "directions" corresponding to the basic functions

$$\sin(t), \quad \cos(t), \quad \sin(2t), \quad \cos(2t), \quad \cdots.$$

For quantum mechanics von Neumann needed a generalization of this construction, obtained by permitting complex numbers (e.g., $2 + 3\sqrt{-1}$) as coefficients multiplying the basic functions.

[p. 63] Heisenberg anecdote: see Macrae, p. 142.

[p. 63] Arthur Wightman assured me that they did read von Neumann in Princeton in the 1940s.

[p. 65] Von Neumann reaches the punchline on p. 325 of his book, von Neumann (1955).

[p. 66] The example of an unstable dynamical system is called "Sinai's billiard" after the Russian mathematical physicist Ya. Sinai; see Sinai (1976).

[p. 67] See Jammer (1974) for a detailed history of the responses and criticisms to von Neumann's theorem. Bell's example is in his first paper "On the Problem of Hidden Variables in Quantum Theory," *Rev. Mod. Phys.* **38** (1966), reprinted in Bell (1987), and his analysis is trenchant. But for the author, only rediscovering Bohm's and Bub's model of a quantum apparatus (see Chapter 18) finally caused the fog to lift.

[p. 68] Other "impossibility theorems" are those of Gleason, *J. Math. & Mech.* **6**, 885 (1957); Jauch and Piron, *Helv. Phys. Acta.* **36**, 827 (1963); Kochen and Specker, *J. Math. & Mech.* **17**, 59 (1967), and there are many more. Although these authors proved interesting theorems about Hilbert spaces, they failed to attain their goal for the same reason as did von Neumann: ignoring the role of the apparatus *in the alternative "hidden variable" theories*. Statisticians, naturally, are the group that understands this point best; see Chapter 12, "Dice Games and Conspiracies," and Chapter 20, "Speculations," for further discussion. These "impossibility theorems" downed a clay target; realism and determinism flew on.

Chapter 9 EPR

[pp. 70–72] This famous exchange appeared in *Physical Review*, **47** and **48**, (1935); reprinted in Wheeler and Zurek. Bohr's letter appeared in *Nature*, **136,** p. 65.

The reader may find it strange—I certainly do—that neither Bohr nor Einstein brought up von Neumann's "impossibility proof" in their debates. But reflecting on von Neumann's membership in another generation, on his profession as a mathematician, and on the date his book appeared (only a few years prior to EPR) makes this lacuna appear less mysterious. We do know that Bohr later held many discussions with von Neumann at Princeton, and we can surmise he was not troubled by von Neumann's conclusions. We also know that, by around 1938, Einstein knew about von Neumann's theorem. Peter Bergmann told Abner Shimony that he, Valentin Bargmann (both assistants of Einstein) and Einstein once discussed it in Einstein's office. On that occasion Einstein took down von Neumann's book, pointed to the additivity assumption, and asked *why we should believe in that*.

[p. 72] Bell expresses bafflement over Bohr: see the appendix to his paper "Bertlmann's Socks and the Nature of Reality," *J. de Physique*, Colloque C2, suppl. au numero 3, Tome **42** (1981), reprinted in Bell (1987). Others who could not understand Bohr included Bohm, Aharanov, and Shimony (see Chapter 13), making his EPR paper virtually a Rorschach test for distinguishing positivists from realists.

[p. 74] Heisenberg's remark about retrodiction is from *The Physical Principles of the Quantum Theory* (1929). As far as I know, Heisenberg did not make the "useless knowledge" objection to EPR either, perhaps because he had adopted Bohr's views after 1927.

[pp. 75–76] Einstein's summing up appeared in *J. Franklin Institute* **221**, no. 3 (1936). In private Einstein was less reserved in his views, referring to Bohr as a mystic (Einstein to Schrödinger, November 1950; see Przibraum (1967) p. 37), who thought of himself as a prophet (see Shankland, R. S., *Am. J. Phys.* **31** [1963], p. 47). Bohr's last sketch: Jammer (1974), p. 120.

Personal Note. The only participant in these great debates from the 1930s I was privileged to meet in person was Dr. Werner Heisenberg, who visited the University of Washington in Seattle in the early 1970s when I was a graduate student there. (He died a few years later.) Although I was one of perhaps two dozen physics students permitted to sit at the feet of the master that afternoon, I did get a question in. I asked: "How did people react to the EPR paper?" The great man thought for a second and then replied, with a smile playing around his eyes, "Many people didn't like it!"

Chapter 10 The Post-War Heresies

[p. 77] Opening quote: from "On The Impossible Pilot Wave," *Found. Phys.* **12** (1982), pp. 989–999, reprinted in Bell (1987). A superb defense of Bohm/de Broglie.

[pp. 79–83] Bohm's papers appeared in *Physical Review* **85** (1952), reprinted in Wheeler and Zurek. The identity of the physicist who debated Bohm has proven impossible to pin down due to conflicting memories of the participants. On later reflection I believe the issue discussed may not be as insignificant as I then thought; the symmetry between position and momentum may be fundamental. For Bohm's mature philosophy, see his book, Bohm (1980). For recent expositions of Bohm's interpretation, see Holland (1993) and Dürr, Goldstein, and Zanghi, *J. Stat. Phys.* **67**, nos. 5/6, 1992.

[p. 81] Bohm's equations for the "physical variables," a positive scalar $\rho(\vec{x}, t)$ and a vector field $\vec{v}(\vec{x}, t)$, have the following form:

$$\vec{v}(\vec{x}, t) = \frac{\nabla S}{m}(\vec{x}, t) \tag{1}$$

($\vec{v}$ is the gradient of another scalar field S);

$$\frac{\partial \rho}{\partial t} + \nabla \cdot (\rho \vec{v}) = 0 \tag{2}$$

$$\frac{\partial S}{\partial t} + \frac{|\nabla S|^2}{2m} + V(\vec{x}) = \frac{h^2}{16\pi^2 m}\left[\frac{\nabla^2 \rho}{\rho} - \frac{1}{2}\frac{|\nabla \rho|^2}{\rho^2}\right]. \tag{3}$$

In these equations m stands for the particle's mass, and h is Planck's constant. Equation 2 is the mass-conservation equation from fluid dynamics, here expressing conservation of probability. Equation 3 with h set equal to zero is the classical Hamilton-Jacobi equation; with h nonzero the right-hand side gives Bohm's "quantum potential." Having solved these equations, the particle's motion is determined by

$$\frac{d\vec{x}}{dt} = \vec{v}(\vec{x}, t). \tag{4}$$

The relation of these variables to Schrödinger's wave function $\psi(\vec{x}, t)$ is

$$\psi(\vec{x}, t) = \sqrt{\rho}\exp[\sqrt{-1}\, S/h]. \tag{5}$$

It must be admitted that eq. (3) is to Schrödinger's lovely linear equation as The Ugly Duchess—immortalized by the 16th-century Flemish painter Quintin Matysys, and by Tenniel for the Alice books—is to the Mona Lisa. But is beauty truth in physics?

[p. 82] For Einstein's reaction ("too cheap"), see the Born-Einstein letters, Born (1971), p. 192; for Rosenfeld's, Jammer (1974), p. 295.

[85–87] For the history of Brownian motion consult Pais' *Subtle Is the Lord...* (1982) and Nelson's *Dynamical Theories of Brownian Motion* (1967). For stochastic mechanics see the latter and Nelson's *Quantum Fluctuations* (1985). For a stochastic-mechanical treatment of EPRB, see W. Faris, *Found. Phys.* **12** (1982), pp. 1–26. Féynes' article appeared in *Z. für Physik* **132** (1952),pp. 81–106, quotations taken from a translation by J. Westwater (unpublished). Bachelier's thesis is reprinted in Cootner (1964). The author thanks Dr. Blaine Walgren for this reference.

[p. 90] Feynman's paper on the "space-time" approach to quantum mechanics (with the "remark in passing" about Brownian motion) is in *Rev. Mod. Phys.* **20**, no. 2 (1948), pp. 367–387; see p. 376. Bohm's similar remark is in *Phys. Rev.* **85**, p. 175 (1952). Wiener's and Siegel's hidden variable theory can be found in Rankin, (1966).

Chapter 11 Bell's Theorem

Bell was profiled by physicist and *New Yorker* staff writer Jeremy Bernstein in *Quantum Profiles* (1991). His collected papers on quantum philosophy are in *Speakable and Unspeakable in Quantum Mechanics* (1987), except for "Against 'measurement" (his final work), a scathing critique of the current situation in quantum physics (see Chapter 18, "Principles").

[p. 92] The popular book Bell read was Max Born's *Natural Philosophy of Cause and Chance*, 1949. Von Neumann's book had not yet been translated from the German.

[p. 96] "44 countries on a grain of rice": see the section on miniature writing in the 1989 *Guinness Book of World Records.*

[p. 98] Bell's theorem appeared in *Physics* **1** no. 3, (1964), pp. 195–200, reprinted in Bell (1987). The story of this journal is interesting in its own right. Its full title was *Physics Physique Fizika* (it accepted papers in three languages), and its subtitle was "An International Journal for Selected Articles Which Deserve the Special Attention of Physicists in All Fields." In an editorial foreword, P. W. Anderson (then at Bell Labs, now at Princeton) and B. T. Matthias (U. of California at San Diego) described their unusual serial as analogous to "a journal of literature and general information, such as *Harper's*," for the benefit of physicists who have stopped reading articles even in closely related subfields, "as perforce most of them have long since... in the other sciences." Writers submit articles to *Harper's*, they noted, partly because they will be paid (and one might also remark that the rejection rate at *Harper's* or the *New Yorker* is at least 99 percent). By contrast, many scientific journals, despite lofty goals and high standards, are essentially vanity presses. "Page charges" at these journals can run as high as a hundred dollars a page. Bell recalled sending his paper to *Physics* at least partly to avoid such charges—he was too embarrassed to ask his American hosts to pay for his unusual submission.

Philip Anderson, who won the Nobel Prize in 1977 for developing the quantum theory of electrons in disordered materials, read Bell's paper and accepted it, partly because he thought it refuted Bohmism.* *Physics Physique Fizika* lasted only four years. It is sad that Anderson's and Matthias's idea did not catch on.

[p. 98] The proof of Bell's theorem given here seems the simplest and is in general circulation among Bell aficionados; I do not know whom to credit. Bell proved his inequality ("Bell's inequality" has since become a generic label for any inequality ruling out local realism) by conventional probability calculus. The angles chosen are not optimal for violating Bell's inequality; with better choices the violation can become $2\sqrt{2}$. Other proofs can be found in Clauser and Horne, *Phys. Rev. D* **10** (1974), p. 526; Wigner, *Am. J. Phys.* **33** (1970), pp. 1005–1009; Stapp, *Phys. Rev. D* **3** (1971), pp. 1303–1320; Eberhard, *Nuovo Cimento B* **38**, (1977), p. 75–80, and **46** (1978), pp. 392–419; Mermin, *Physics Today* (April 1985), p. 38, and there are many others (peruse the back issues of *Found. Phys.*, for instance).

Bell's theorem entered popular culture with Gary Zukav's *The Dancing Wu-Li Masters* (1979), where it is called, somewhat hyperbolically, "the most important single work, perhaps, in the history of physics." Unfortunately, Zukav continues, it is also "indecipherable to the non-mathematician."

Chapter 12 Dice Games and Conspiracies

Concerning relativity theory, I recommend *General Relativity: From A to B* by Robert Geroch (1978), from whom I learned Einstein's theory in college. In this excellent book with a modest title, the reader will learn about the special and

*Private communication to the author.

general theory from a man who collaborated with Steven Hawking and Roger Penrose on proving that black holes are inevitable in most cosmologies. Also recommended is *Relativity for the Million* (1962) by Martin Gardner; no one does this kind popular science writing better. *The Principle of Relativity*, by H. A. Lorentz et al., is the little book of translations mentioned in the Introduction; Minkowski's 1908 lecture is included.

For experimental tests of special relativity, which have attained an accuracy of 70 parts per million in direct tests of the constancy of the speed of light, see *Science* **30** (November 1990), pp. 1207–1208. As if to fulfill the journalistic requirement of "balanced treatment," *Science* included a report on two heretics who are trying to disprove relativity. (Note: No heretic mentioned in this book tried to *disprove* quantum mechanics.)

Just for the fun of it, take a look also at the chapter on relativity in one of the standard textbooks, say Halliday and Resnick (first edition, 1960; latest 1992). From their Hollywood version of the birth of relativity, one cannot help visualizing gray-bearded scientists in seminar lamenting their aether-drift experiments, when suddenly the young Einstein pops up from the audience with the shout: "I have a hypothesis!" The hypothesis contradicts what everyone has believed about velocity or time since grade school, yet the unknown proposer is lauded as a genius and his miraculous new "axiom" wins the day. The truth is that people who do that are called "cranks" and are shown quickly to the door. Einstein had very plausible principles (see Chapter 18 "Principles") as well as convincing arguments backing up his proposals. This kind of presentation suggests that progress in science comes from wild hypothesizing rather than from painstaking analysis; perhaps the recent "cold fusion" and "fifth force" episodes can be partly blamed on such mythologizing. Of course, the standard textbook presentation of quantum mechanics is even worse than for relativity—but for this the writers cannot be held entirely responsible!

[p. 101] The Einstein anecdotes are from Pais, *Subtle Is the Lord...* (1982).

[p.106] The ridiculous scenario is constructed by starting with a tachyon emitted by the first car at rest and making a Lorentz transformation to a new coordinate system in which the car is moving to the left and the tachyon is moving to the right and down (e.g., "backward in time"). For the second car, reflect this construction and move it down a bit; see Figure 21. Note that all time statements refer to a single reference frame.

[p. 108] The analyst for the Navy: see Mermin, *Physics Today* (April 1985).

[p. 109] References on local causality are: J. F. Clauser and M. Horne, *Phys. Rev. D* **10** (1974), pp. 526–35; Bell's articles "The Theory of Local Beables" and "Bertlmann's Socks and the Nature of Reality" in Bell (1987); "An Exchange on Local Beables," *Epistemological Letters* (Feb. 1977), reprinted in *Dialectica* **39** (1985); and J. F. Clauser and A. Shimony, *Rep. Prog. Mod. Phys.* **41** (1978), pp. 1881–1927.

[pp. 111–112] Concerning Nelson's distinction between active and passive locality, it is a remarkable fact that, as of 1994, the only theory known to satisfy the one but not the other is quantum mechanics. Of course, quantum theory

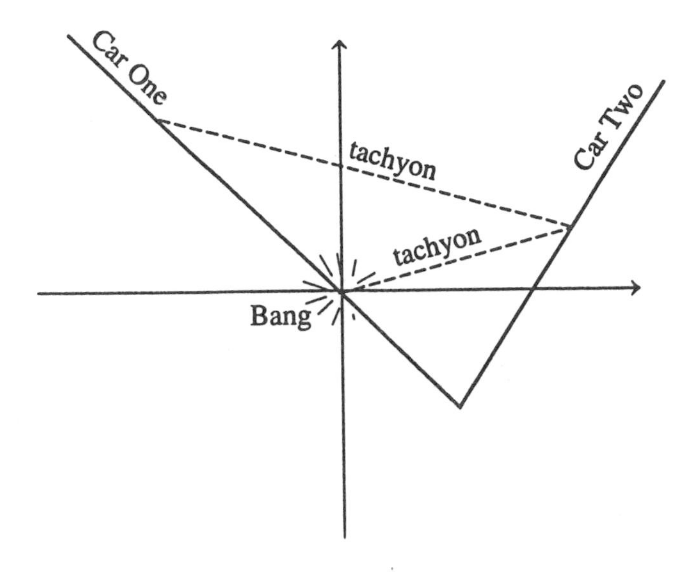

Figure 21

accomplishes the trick by refusing to provide those missing variables (that is, it does not satisfy a realism principle). Efforts are underway to find a realistic, stochastic field theory that violates passive locality, especially by Nelson and his colleagues, but it is a very difficult business.

[p. 113] Nelson's version of Bell's Theorem is in S. Albeverio et al. (1986). The argument that perfect correlations plus independent stochasticity implies determinism (see the Appendix) was made by P. Suppes and M. Zanotti; see Suppes (1976). What I have called "passive locality" was first described by J. Jarrett, *Nous* **18** (1984), pp. 569–589, who called it the "completeness condition"; it was also discussed by Shimony under the name "uncontrollable nonlocality" or "outcome independence." Shimony also coined a more picturesque phrase for this kind of nonlocality—less than action at a distance but more than mere correlation: "passion at a distance." See Shimony (1993) and references therein. Michael Redhead has distinguished *five* forms of nonlocal influence (and three major interpretations of quantum mechanics); see Redhead (1987).

Chapter 13 Testing Bell

This chapter was based partly on interviews with the principals, and on unpublished material kindly provided to the author. Published articles can be referenced from the review article of Clauser and Shimony, *Rev. Prog. Phys.* **41** (1978), pp. 1881–1927. The first photon correlation experiment was reported by E. Brannen, F. Hunt, R. Adlington, and R. Nichols in *Nature* **175** (1955), p. 810. Another test of spin correlations was performed in 1976 by Lamehi-Rachti and

Mittig, *Phys. Rev. D* **14** no. 10, p. 2543, using low-energy proton-proton scattering; the quantum formula held.

[p. 117] Hans Freistadts' review article appears in *Nuovo Cimento*, **5**, Supp. Vol. 1–70 (1957).

[p. 117] In technical language, Schrödinger had proposed that when two systems become sufficiently separated, the *S*-state wave function falls into a product state without correlations; see the Appendix.

As I write, John Clauser and Stuart Freedman are research physicists at the University of California in Berkeley. They are well known for their work in atomic and nuclear physics, respectively; Clauser particularly for work on atomic interferometry (he holds a patent on the single-atom interferometer), and Freedman for, as he puts it, "a long series of null experiments." (By his count Freedman has performed 23 experiments each ruling out some kind of exotic particle or effect—most recently, a conjectured massive neutrino—and is beginning to wonder if his thesis work with Clauser started some kind of trend.) Abner Shimony is professor of physics and philosophy at Boston University, has written extensively on quantum nonlocality and realism, and is president-elect of the Philosophy of Science Association. Michael Horne is professor of physics at Stonehill College and continues to propose experiments in many-particle quantum mechanics. Richard Holt is a physics professor at the University of Western Ontario and performs experiments in atomic physics.

Chapter 14 Loopholes

[p. 129] Mermin's article appeared in *Physics Today* (April 1985), p. 38. See also the Letters column in November 1985, p. 9. Mermin is a regular contributor to *Physics Today* and writes frequently about quantum conundrums.

[p. 130] About the 70,000 journals: the number listed in Bowker-Ulrich's database as of December 7, 1990, was 74,000; see *Science* on that date, p. 1331. A survey of the top 4,500 journals undertaken for *Science* revealed that 55 percent of scientific papers are never cited by other authors, serving only to increase the author's "page count."

[p. 134] The first and third loopholes were first described by Clauser and Horne, *Phys. Rev. D* **10** (1974), p. 526; see also Clauser and Shimony's 1978 review (*op. cit.*, Chapter 13) and T. K. Lo and A. S. Shimony, *Phys. Rev. A* **23**, 6 (1981).

[p. 134] Aspect's experiment: see A. Aspect, J. Dalibard, and G. Roger, *Phys. Rev. Lett.* **49** (1982), p. 1804. That such rapid switching should be incorporated in the experiment was first mentioned in print by Shimony in 1971; see *Foundations of Quantum Mechanics*, B. d'Espagnat, ed., who attributes the idea to Clauser. Aspect's water baths did not entirely close the communication loophole, however, since the vibrations in the baths are periodic, and hence the vibration phases might be exploited by some hidden communication scheme.

See also E. Santos, *Phys. Rev. Lett.* **66** 11, pp. 1388–1390, and Errata **66**, no. 24, pp. 3227, a comment/rejoinder **68**, no. 17, pp. 2701–2703, and related work: S. Caser, *Phys. Lett.* **121** (1987), p. 131.

Chapter 15 The Impossible Observed

This chapter is based on interviews with Hans Dehmelt, who kindly provided a tour of his lab, and with his collaborators. I thank Dr. Werner Neuhauser for providing additional material.

[p. 137] For the visible barium ion, see *Phys. Rev. A*, **22**, no. 3 (1980), pp. 1137–1140, and the color snapshot from Dehmelt's lab reproduced in the *Phys. Scripta* article. (Unfortunately, "Astrid," Dehmelt's pet barium ion, was not available for viewing the day I visited his lab, but I accept that others have seen it. I haven't seen Antarctica either.)

[p. 137] Schrödinger quote: *Brit. J. Philos. Sci.* **III**, no. 11 (Nov. 1952).

[p. 138] R. H. Dicke and H. A. Tolhoek earlier (in 1956) attempted to measure the electron's magnetic moment, and W. H. Louisell, R. W. Pidd, and H. R. Crane first succeeded (accuracy 3 percent); see Van Dyck et al., *op cit.* (1986).

[p. 138] For a report on the 1989 Nobel Prize in Physics, see *Science*. 20, (Oct. 1989), p. 327. Dehmelt's Nobel lecture was published in *Rev. Mod. Phys.* **62**, no. 3 (1990), pp. 525–530. (The values for the electron's g-factor, which may change over the years with refinements in experimental and numerical technique, are taken from that lecture.) Other interesting articles include Van Dyck et al., *Phys. Rev. D* **34**, no. 3, (1986), pp. 722–734; H. Dehmelt, *Phys. Scripta* **T22** (1988), pp. 102–110, and *Z. Phys. D.* **10** (1988), pp. 127–134.

[p. 144] For quantum jumps observed, see W. Neuhauser and Th. Sauter, *Comments At. Mol. Phys* **21**, no. 2 (1988), pp. 83–95, and Dehmelt's papers above.

[p. 145] For Kinoshita's heroic calculations, see *Proc. of the Conference on Precision Electromagnetic Measurements*, Gaitherburg, 1986. The error is less than the experimental uncertainty in the fine structure constant since the latter enters into the calculations raised to a power of four or higher. Numeric evaluation of integrals accounts for the remaining error.

[p. 144] Kerson Huang's paper on electron spin is in *Am. J. Phys.* **20** (1952), p. 479. On the history of spin, see Jammer (1966), pp. 147–53.

[p. 145] Bohr on spin is from an unpublished English-language manuscript dated 1929 entitled "The Magnetic Electron," in the Niels Bohr Archives, microfilm MSS, no. 12. Pauli's article from the Sixth Solvay Conference can be found in *Le Magnetism. Rapports et Discussions de Sixieme Conseil de Physique . . .*, Gauthiers, Paris (1932), in French, or in the Bohr Archives in English. Mott's article was in *Proc. Roy. Soc. A* **124** (1929), p. 440; see also pp. 214–219 of Mott and Massey (1965), reprinted on pp. 701–706 of Wheeler and Zurek. Rosenfeld's recollection is quoted in Wheeler and Zurek, p. 699.

Chapter 16 Paradoxes

[p. 148] Feynman quote: Feynman (1965), p. 129.

[p. 148] **1.** "Spooky action at a distance"

For Einstein's use of this phrase, see the letter from Einstein to Born dated March 1947, Born (1971). Another illustration of *spukhafte Fernwirkungen* is in EPRB, if one invokes the collapse of the wave function after a measurement at one of the analyzers. Mathematically, this collapse replaces the wave function by a "statistical mixture" (diagonal statistical operator); if the real state of affairs is described by this operator, action at a distance must be involved.

[p. 149] **2.** Schrödinger's cat

The celebrated cat made its appearance in *Naturwissenschaften*, Vol. 23, in three parts (the paper, not the cat); a translation into English appeared in *Proc. Am. Phys. Soc.* **124** (1980), p. 323, reprinted in Wheeler and Zurek. Quotations are from this translation. Related English-language papers (Schrödinger was in Cambridge at the time) are in *Proc. Cambridge Phil. Soc.*, **31** and **32**.

Somehow the cat does not always serve Schrödinger's original purpose of barring ill-conceived speculation. In an otherwise praiseworthy piece in *Discover*, (Nov. 1990), p. 64, the writer asserted that if certain proposed experiments work out as planned, "physicists may have to accept the idea that a cat could be both dead and alive. . . ." This is no more likely than that creditors will one day discover that Donald Trump is both rich and broke. If the proposed experiments with SQUIDS (Superconducting QUantum Interference Devices) exhibit macroscopic quantum effects as hoped, the experimenters will not come out for smeared-out felines but for a new description of UP-CAM, wave collapse, or a better theory generally.

[p. 152] **3.** Uncertainty over the uncertainty principle

Other interpretations of Heisenberg's laws can be found in Jammer (1974), in back copies of *Phys. Rev.*, *Found. Phys.*, and in many other journals.

Have experiments demonstrated Heisenberg's laws? Unfortunately, it is difficult to prove a negative; as Jammer (1974, p. 81) remarked: "Rarely in the history of physics has there been a principle of such universal importance with so few credentials of experimental tests."

[p. 156] Soft pillow: letter of Einstein to Schrödinger, May 1928; see Przibraum (1967), p. 31.

Late in life, Heisenberg adopted a more metaphysical interpretation of his famous relations. (See his book *Physics and Philosophy*, published in 1962.) He argued that "potentiality," thought of as an alternate state lying between possibility and actuality, might explain the situation. After a position measurement, the particle's momentum might be reduced to a mere potentiality. This strikes me as trying to solve a physical puzzle by waving a linguistic wand, a trick attempted too often by physicists in this century.

[p. 156] **4.** Quantum probability, or: what physicists do not speak about

Feynman's 1951 lecture appeared in *Proc. 2nd Berkeley Symp. on Math. Stat.*, University of California Press (1951).

[p. 159] **5.** Time and the statistics of particles and people

The mathematical statement that "time is not an observable" in quantum mechanics is that there is no self-adjoint operator representing the time of any event. In particular, there is no operator "T" obeying the Born-Jordan commutation relation with the Hamiltonian H (representing the energy), an observation made originally by Pauli; see Jammer (1974) p. 41. As the Born-Jordan relation is used in the standard proofs of Heisenberg's first law, the usual textbook account of the "uncertainty relations of quantum mechanics" contained a bit of fibbing for many years. The best treatment of the time-energy uncertainty relation I know of is in Holevo (1982), a good book expounding a statistical interpretation of quantum mechanics. By necessity, Holevo is forced to go outside the usual mathematical framework of quantum theory; his results therefore may or may not be acceptable to other physicists.

[p. 162] **6.** Can a graduate student collapse the wave packet?

The paradox of Wigner's friend can be found in his paper on the mind-body question contained in *The Scientist Speculates,* I. J. Good, ed. (1961), reprinted in *Symmetries and Reflections*, by E. Wigner (1967), and Wheeler and Zurek (1983).

[p. 165] "Infamous boundary" is from Bell's essay "Introduction to the Hidden-Variable Question," reprinted in *Speakable and Unspeakable in Quantum Mechanics* (1987), p. 35.

[p. 166] **7.** "A watched pot never boils"

The "watched-pot-never-boils effect" has been rediscovered many times; see Misra and Sudarshan, *J. Math Phys.* **18**, no. 4 (1977), pp. 756–763, and references therein. The mathematics of the effect is related to the famous limit theorem known to calculus students:

$$\lim_{n \to \infty} \left[1 - \frac{T}{n}\right]^n = e^{-T}.$$

Here n stands for the number of "peeks," T/n represents the change in the probability in that small time interval, and e^{-T} gives the classical probability that the particle is still in the box after a time T. In quantum mechanics the fraction T/n gets replaced by T/n^2, and now the limit is 1. Exercise for the quantum mechanician: exploit the self-adjointness of the Hamiltonian to show that

$$P[\text{no transition in a time } T] = 1 + O(T^2)$$

as $T \to 0$, for any quantum transition.

[p. 167] A concrete occurrence of the "watched cat never dies" paradox came in Hans Dehmelt's experiments described in Chapter 15. After trapping an electron and arranging to observe its spin-flips on a cathode-ray tube, it occurred to Dehmelt that, due to the effect of watching, he shouldn't be able to see any. Dehmelt's resolution of the paradox is that any observation must take a definite amount of time, so "continuous watching" is impossible.

[p. 169] Protons under observation: Horowitz's and Katznelson's paper appeared in *Phys. Rev. Lett.* **50**, no. 16 (1983), p. 1184, and the replies in the same journal, **51**, no. 1 (1983). I thank Arthur Wightman for bringing this wonderful episode to my attention.

[p. 170] Bollinger et al. experiment: see *Science* (Nov. 17, 1989), p. 888.

Chapter 17 Philosophies

[p. 172] Professional philosophers prefer, in place of my naïve realism, either "critical" or "representational" realism; see the article "Realism" in *The Encyclopedia of Philosophy*, Paul Edwards, ed. The reason is historical, the former being reserved for an archaic doctrine of, for example, sensed greenness residing in grass, as opposed to the modern view that sensation arises from interactions of the sense organs with intermediates (light) or the thing sensed. However, in my time the standard insult was always "You're a naïve realist!" so I won't opt for the less pejorative phrase (which would cause my friends to think I was putting on airs).

[p. 176] Schrödinger on naïve realism: from the cat paper (*op cit.* Chapter 16).

[p. 177] Corpuscles: see Whittaker (1951), pp. 31–32.

[p. 180] Light and Life: *Nature* **131** (1933), pp. 421–457.

[pp. 180–183] On Delbrück: I profited greatly from the article by Lily E. Kay in *J. Hist. Bio.* **18**, no. 2 (1985), pp. 207–246; the articles in Cairns et al. (1966), see especially "Waiting for the Paradox" by Gunther Stent and Delbrück's 1949 address, pp. 22. See also Delbrück's article "A Physicist's Renewed Look at Biology: Twenty Years Later," *Science* **168** (1970), pp. 1312–1315.

[p. 183] Watson and Crick influenced by Schrödinger: see Moore (1989), pp. 403–404. Muller quotation: Moore (1989), p. 402.

[p. 184] *What Is Life?*: Schrödinger (1944). Stent (*op cit.*) refers to the title as a colossal piece of nerve.

[p. 185] Bohr's references to psychology: *Nature* **121** (1928), p. 580–590; *Die Naturwissenshaften* **17** (1929), pp. 483–486; the first reprinted in Wheeler and Zurek.

[pp. 186–187] The Bohr-James connection: see Jammer (1966), Chapter Seven, and James (1890). It is known that Bohr had read James ("I thought he was most wonderful") and that when questioned about the matter by T. S. Kuhn and Aage Peterson, Bohr even mentioned the "Stream of Thought" passage; see the interview of November 1962, with T. S. Kuhn et al., in the Bohr Archives. Also quoted in Holton (1973), p. 137.

[p. 186] Lucie: Jammer (1966), p. 350; James (1890), pp. 202–213; Janet (1889), pp. 276– 277.

Jerome Bruner, the educational theorist, told of meeting a physicist named "Mr. Baker" (Bohr's cover name) in 1944 at a friend's home in Washington, and hearing a different account of the origin of complementarity. Bruner recalled Bohr telling him how he had first thought of complementarity in connection with having to discipline his son. "Could he (Bruner recalled Bohr saying) *know* his son *simultaneously* both in the light of love and in the light of justice? Were these not mutually non-convertible ways of knowing?" (Bruner, 1971, p. xiii.) I thank Dr. Matthew Howard for bringing this reference to my attention.

[p. 188] Heisenberg quotation: from his article "Quantum Theory and Its Interpretations" in Rozental (1967), p. 107.

Chapter 18 Principles

[p. 190] Einstein's principles: from Lorentz et al. (1923). For the Hilbert-Einstein episode, see Pais (1982) and J. Earman and C. Glymour in *Archiv. His. Exact Sciences* **19** (1978), pp. 291–308. A fair judgment might be: it was Einstein's theory, but the Einstein-Hilbert equations of gravity.

[p 193] "Against 'measurement'" was based on a talk Bell gave in Erice, Sicily, in August 1989. See A. Miller (ed.), *62 Years of Uncertainty* (1990), p. 19, reprinted in *Physics World* (August 1990). N. G. van Kampen (one of the three authors mentioned by Bell) angrily replied in that journal in October; Rudolf Peierls (Bell's thesis advisor many years before) in January 1991; and Kurt Gottfried (another author) in October 1991. Bell's criticisms whizzed by these establishment physicists, who repeat old arguments as if unable to hear the dead physicist's voice. For example, Peierls claimed "it is easy to give an acceptable account" of quantum measurements (meaning an account FAPP), and asserted that the wave function represents merely "our knowledge" of the particle. When criticized by a letter writer (May 1991) for ignoring Bell's concerns, he replied "concepts like 'reality' cannot be given any meaning at the quantum level." That may put deluded realists in their place, but it misses Bell's point that physicists never found a precise account of atoms.

[p 195] Quantum logic: see Jammer (1974). Quantum probability: see Accardi and Weldenfels (eds.) (1989). Many worlds: see H. Everett, *Rev. Mod. Phys.* **29** (1957), pp. 454–462, reprinted in Wheeler and Zurek. Another remark about these theories is that, when one gets to the details, one discovers that assumptions no more plausible than those of conventional theory are needed to derive a Hilbert space.

[p 196] Wallstrom first published his observations in *Found. Phys. Lett.* **5**, no. 2 (1989), p. 113, a journal so obscure that he might have done better in *Stamp Collecting News*. He wrote a more comprehensive account for *Phys. Rev.*, **49**, no. 3 (1994), pp. 1613–1617. While preparing the latter, Wallstrom discovered that his discovery had been made earlier, by Takehiko Takabayasi from Nagoya, Japan, in a paper on the hydrodynamical interpretation that appeared in—you guessed it—1952. (See *Prog. Theor. Phys.* **8** (1952), p. 143.) Although Takabayasi did not quite make the connection with topology, he

noted the line-integral condition

$$m \oint_{\gamma} \vec{v} \cdot d\vec{l} = nh$$

(where m denotes the particle's mass, v the velocity, h Planck's constant, n an integer, and the line-integral is taken around any closed curve γ not intersecting the zero-set of the probability density) and remarked

> At any rate we are led to a new postulate... which is so to speak the 'quantum condition' for fluidal motion and of *ad hoc* and compromising character for our formulation, just as [Bohr's angular momentum condition of Chapter 2] was for the old quantum theory.

Wallstrom's observation was that it is equally compromising for stochastic mechanics.

[p. 198] Cartan's book: E. Cartan (1931). Axioms for quantum geometry: manuscript of the author (1991, unpublished).

[p. 200] Wootters' information principle appears in A. Marlow (ed.) *Quantum Theory and Gravitation* (1980), pp.13–26; see also *Phys. Rev.* **23** (1981), pp. 357–362. The observation that minimizing the error in estimating a parameter leads to the rule "probability equals the square of a projection" dates back to a paper of R. A. Fisher, one of the founders of modern statistics; see *Proc. Royal Soc. Edinburgh* **62** (1922), pp. 321–341.

[p. 202] The square root of minus one: try this exercise. Take a finite-dimensional quantum system (e.g., an N-level atom) with Schrödinger equation

$$\sqrt{-1}\frac{\partial}{\partial t}\Psi_i = \sum_{j=1}^{N} H_{i,j}\Psi_j,$$

where $H_{i,j}$, the quantum Hamiltonian, is a Hermitian matrix. Let

$$\Psi_i = q_i + \sqrt{-1}p_i,$$

for $i = 1, \ldots, N$ (never mind whether q_i and p_i mean anything physically), separate real and imaginary parts in Schrödinger's equation, and identify what you get. Hint: remember that

$$\mathcal{H} \equiv \langle\psi|H|\psi\rangle = \sum_{i,j=1}^{N} \bar{\Psi}_i H_{i,j} \Psi_j$$

is a constant of the motion. Harder exercise: identify all such (real) evolution equations.

[p. 202] Bohm and Bub: *Rev. Mod. Phys.* **38** (1966), p. 447. C. Papaliolios, *Phys. Rev. Lett.* **18**, no. 15 (1967), tested one consequence of the model, with negative results. K. Hepp, *Helv. Acta.* **45** (1972), p. 237, proposed that *in the* *limit of infinitely large systems* quantum mechanics can resolve a dichotomy. See Bell (1987), paper six, for a criticism. When Arthur Wightman advertised Hepp's result in a seminar at Princeton, Eugene Wigner is said to have remarked, "You're a great man, Arthur, but you are not infinite."

[p. 204] The GRW scheme: see *Phys. Rev. D* **34**, no. 2 (1986), p. 470, and Miller, *62 Years of Uncertainty* (1990).

Solution to the exercise: The equations are the classical ones associated with the Hamiltonian $\mathcal{H}$. The dynamical laws of this type are precisely the classical, linear Hamiltonian flows that commute as flows with the maximally symmetric case ($H_{i,j} = \delta_{i,j}$).

Chapter 19 Opinions

[p. 205] No quantum world: A. Peterson in French and Kennedy (1985).

[p. 206] Feynman on the two-slit experiment: it begins on p. 129 of Feynman (1965) and continues on to its aggressive conclusion on p. 148. The reader might enjoy comparing and contrasting this passionate essay with Bell's last paper (see Notes to Chapter 18), although since Bell wrote for his fellow physicists and Feynman for students, Feynman's is more accessible.

[p. 208] Bell quotation: from "On the Impossible Pilot Wave," reprinted in Bell (1987).

[p. 209] Bohr quotation (concerning complementarity): from *Nature* **121** (1928), p. 580, reprinted in Wheeler and Zurek.

[p. 210] For quantum field theory, see Streater and Wightman, *PCT, Spin, Statistics and All That* (1964).

[p. 212] Griffiths' consistent histories is in *J. Stat. Phys.* **36**, nos. 1/2 (1984).

[p. 214] For "decoherence," see Gell-Mann and Hartle, in Zurek (1990), or Hartle in Coleman et al. (1991). Zurek wrote a semipopular account in *Physics Today*, (Oct. 1991), pp. 36–44, and critics and supporters thrashed it out in the Letters column for April 1993.

Chapter 20 Speculations

[pp. 216–217] The "hidden assumption" in the impossibility theorems is the following. The theorems compare quantum mechanics to a theory of a (possibly multidimensional) hidden variable, call it λ, for which there is a universal probability law $P(d\lambda)$. To each observable A, B, ... (Hermitian matrices in quantum mechanics), there is assumed to be a corresponding random variable $f_A(\lambda)$, $f_B(\lambda)$, ... After making suitable additional assumptions about these random functions or their expected values, a contradiction is derived with some prediction of quantum mechanics (references in Chapter 8). But a statistician would object at the outset. Since different A's may represent different experiments, possibly requiring different apparatus, the statistician's instinct is to write $P_{A,B,...}(d\lambda)$ for the probability law and $f(\lambda)$ for the outcome variable, as in the model in the text.

[p. 221] Some philosophers expressed the opinion that the "contextual/noncontextual" distinction has to do with being versus observing, with "is" as opposed to "is found to be" (see Jammer, 1974). But I side with the statistician wrinkling her brows about these speculations. The experiments involve different situations, she thinks; metaphysics is not required.

[p. 221] Lumps and wires: the English mathematician Roger Penrose developed a similar idea for combining "spins"; see his article in Hawking and Israel (1987). He derived a discrete arithmetic of these spins, and a "principle of indifference" to explain the randomness, which reproduces all the quantum predictions for spin measurements. However, recovering the continuous four-dimensional space-time manifold from his discrete construction proved elusive.

In his Princeton lecture, Bohm suggested something like the first speculation, if I understood him correctly.

Bibliography

Andrade, E. N. da Costa, *Rutherford and the Nature of the Atom*, Doubleday, New York, 1964.

Accardi, L., and Waldenfels, W. (eds.), *Quantum Probability and Applications IV*, Proceedings of the Year of Quantum Probability, Rome 1987, Springer, Berlin, 1989.

Albeverio, S., Casati, G., and Merlini, D. (eds.), *Stochastic Processes in Classical and Quantum Systems*, Proc. Ascona, Switzerland, 1985, *Lecture Notes in Physics*, **262**, Springer, Berlin, 1986.

Bell, J. S., *Speakable and Unspeakable in Quantum Mechanics*, Cambridge University Press, Cambridge, 1987.

Bernstein, J., *Quantum Profiles*, Princeton University Press, Princeton, NJ, 1991.

Blackmore, J. T., *Mach: His Work, Life and Influence*, University of California Press, Berkeley, 1972.

Bohm, D., *Wholeness and the Implicate Order*, Routledge & K. Paul, London, 1980.

Born, M. (ed.), *The Born-Einstein Letters*, Macmillon, London, 1971.

Broda, E., *Ludwig Boltzmann: Man, Physicist, Philosopher*, Ox Bow Press, Woodbridge, CT, 1983.

Bruner, J. S., *The Relevance of Education*, Norton, New York, 1971.

Cairns, J., Stent, G., and Watson, J. (eds), *Phage and the Origins of Molecular Biology*, Cold Spring Harbor Laboratory of Quantitative Biology, 1966.

Cartin, E., *Lecons sur la Geometrie Projective Complex*, Gauthier-Villars, Paris, 1931.

Cassidy, D. C., *Uncertainty: The Life and Times of Werner Heisenberg*, W. H. Freeman & Co., New York, 1992.

Cohen, R. S., and Seeger, R. (eds.), *Ernst Mach: Physicist and Philosopher*, Boston Studies in the Philosophy of Science Vol. VI, D. Riedel, Dordrecht, Holland, 1970.

Coleman, S. et al. (eds.), *Quantum Cosmology and Baby Universes*, World Scientific, Singapore, 1991.

Cootner, P. H. (ed.), *The Random Character of Stock Market Prices*, MIT Press, Cambridge, 1964.

Courant, R., and Hilbert, D., *Methoden der Mathematischen Physik*, Springer, Berlin, 1924.

Einstein, A., *Investigation in the Theory of Brownian Motion*, Methuen and Co., 1926; also Dover Publications, New York, 1956.

Frank, P., *Einstein, His Life and Times*, Knopf, New York, 1947.

Feynman, R., *The Character of Physical Law*, M.I.T. Press, Cambridge, 1967. (First published by the British Broadcasting Corporation, London, 1965.)

Feynman, R., Leighton, R., and Sands, M., *The Feynman Lectures on Physics*, Addison-Wesley, Reading, MA, 1963.

Feynman, R., and Hibbs, A. R., *Quantum Mechanics and Path Integrals*, McGraw-Hill, New York, 1965.

French, A. P., and Kennedy, P. J. (eds.), *Niels Bohr: A Centenary Volume*, Harvard University Press, Cambridge, MA, 1985.

Gardner, M., *Relativity for the Million*, Macmillan, New York, 1962.

Geroch, R., *General Relativity from A to B*, University of Chicago Press, Chicago, 1978.

Good, I. J., *The Scientist Speculates*, Heinemann, London, 1961.

Halliday, D., and Resnick, R., *Physics*, Vols. I and II, Wiley, New York, 1992.

Hawking, S. W., and Israel, W. (eds.), *300 Years of Gravity*, Cambridge University Press, Cambridge, U. K., 1987.

Heisenberg, E., *Inner Exile: Recollections of a Life with Werner Heisenberg*, Birkhäuser, Stuttgart, 1980.

Heisenberg, W., *The Physical Principles of the Quantum Theory*, University of Chicago Press, Chicago, 1929.

— *Physics and Philosophy. The Revolution in Modern Science*, Harper and Row, New York, 1962.

— *Physics and Beyond. Encounters and Conversations*, Harper and Row, New York, 1972.

— *The Physicist's Conception of Nature*, Greenwood Press, Westport, CT, 1970.

Heims, S. J., *John von Neumann and Norbert Weiner: From Mathematics to the Technologies of Life and Death*, M.I.T. Press, Cambridge, MA, 1980.

Hermann, A., *Werner Heisenberg 1901–1976*, Inter Nationes, Godesberg, 1976.

Holland, P. R., *The Quantum Theory of Motion: An Account of the de Broglie-Bohm Casual Interpretation of Quantum Mechanics*, Cambridge University Press, Cambridge, U. K., 1993.

Holevo, A. S., *Probabilistic and Statistical Aspects of Quantum Theory*, North-Holland, Amsterdam, 1982.

Holton, G., *Thematic Origins of Scientific Thought: Kepler to Einstein*, Harvard University Press, Cambridge, 1973.

Instituts Solvay, *Fundamental Problems in Elementary Particle Physics*, Proc. 14th Conf., Interscience, New York, 1968.

James, W., *The Principles of Psychology*, H. Holt and Co., New York, 1890.

Jammer, M., *The Conceptual Development of Quantum Mechanics*, McGraw-Hill, New York, 1966.

— *The Philosophy of Quantum Mechanics*, John Wiley & Sons, New York, 1974.

Janet, P., *L'Automatisme Psychologique*, 10th edition, Alcan, Paris, 1889; 1930.

Kuhn, T., *Black-Body Theory and the Quantum Discontinuity, 1894–1912*, Clarendon Press, Oxford, 1978.

Lorentz, H. A. et al., *The Principle of Relativity*, Methuen and Co., 1923; also Dover Publications, New York, 1952.

Ludwig, G., *Wave Mechanics*, Pergamon Press, Oxford, 1968.

Macrae, N., *John von Neumann*, Pantheon, New York, 1992.

Marlow, A. (ed.), *Quantum Theory and Gravitation*, Academic Press, New York, 1980.

Messiah, A., *Quantum Mechanics*, Vols. I and II, North-Holland, Amsterdam, and John Wiley & Sons, New York, 1958.

Miller, A. (ed.), *62 Years of Uncertainty*, Plenum, New York, 1990.

Moore, W., *Schrödinger, Life and Thought*, Cambridge University Press, Cambridge, U. K., 1989.

Mott, N., and Massey, H., *The Theory of Atomic Collisions*, Clarendon Press, Oxford, 1965.

Nelson, E., *Dynamical Theories of Brownian Motion*, Princeton University Press, Princeton, NJ, 1967.

— *Quantum Fluctuations*, Princeton University Press, Princeton, NJ, 1985.

Pais, A., *'Subtle Is the Lord... ' The Science and the Life of Albert Einstein*, Oxford University Press, Oxford and New York, 1982.

— *Inward Bound. Of Matter and Forces in the Physical World*, Oxford University Press, Oxford and New York, 1986.

— *Neils Bohr's Times, In Physics, Philosophy and Polity*, Clarendon Press, Oxford, 1991.

Price, W., and Chissick, S. (eds.), *The Uncertainty Principle and Foundations of Quantum Mechanics: A Fifty Years' Survey*, John Wiley & Sons, New York, 1977.

Przibram, K. (ed.), *Letters on Wave Mechanics*, Philosophical Library, New York, 1967.

Rankin, B. (ed.), *Differential Space, Quantum Systems, and Prediction*, M.I.T. Press, Cambridge, MA, 1966.

Redhead, M., *Incompleteness, Nonlocality and Realism*, Clarendon Press, Oxford, 1987.

Rozental, S. (ed.), *Niels Bohr: His Life and Work as Seen by His Colleagues*, North-Holland, Amsterdam, 1967.

Schrödinger, E., *What Is Life? The Physical Aspects of the Living Cell*, Cambridge University Press, Cambridge, U. K., 1944.

Shimony, A., *Search for a Naturalistic World View*, Vols. I and II, Cambridge University Press, Cambridge, U. K., 1993.

Sinai, Ya., *Introduction to Ergodic Theory*, Princeton University Press, Princeton, NJ, 1976.

Streater, R., and Wightman, A., *PCT, Spin, Statistics and All That*, Benjamin, New York, 1964.

Suppes, P., *Logic and Probability in Quantum Mechanics*, Reidel, Dordrecht, 1976.

ter Haar, D., *The Old Quantum Theory*, Pergamon Press, Oxford, 1967.

Van Der Waerden, B. L. (ed.), *Sources of Quantum Mechanics*, Dover, New York, 1967.

Wheeler, J. A., and Zurek, W. H. (eds.), *Quantum Theory and Measurement*, Princeton University Press, Princeton, NJ, 1983.

Whittaker, E. T., *History of the Theories of Aether and Electricity*, Thomas Nelson, London, 1951.

Wigner, E., *Symmetries and Reflections*, Indiana Universitiy Press, Bloomington, 1967.

Von Neumann, J., *Mathematical Foundations of Quantum Mechanics*; trans. R. T. Beyer, Princeton University Press, Princeton, NJ, 1955.

Zurek, W. (ed.), *Complexity, Entropy, and the Physics of information*, Addison-Wesley, Redwood City, CA, 1990.

Index

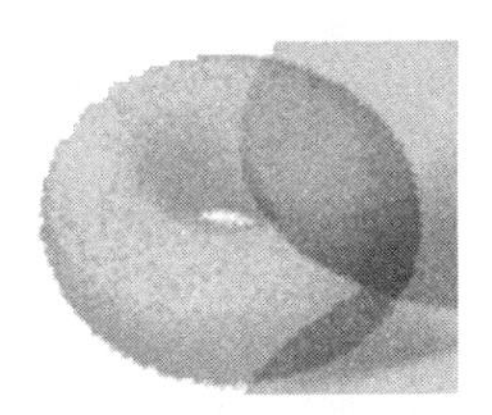